Blood on Steel

Witness to History

Peter Charles Hoffer and Williamjames Hull Hoffer, Series Editors

ALSO IN THE SERIES:

Williamjames Hull Hoffer, *The Caning of Charles Sumner: Honor, Idealism, and the Origins of the Civil War*

Tim Lehman, *Bloodshed at Little Bighorn: Sitting Bull, Custer, and the Destinies of Nations*

Daniel R. Mandell, *King Philip's War: Colonial Expansion, Native Resistance, and the End of Indian Sovereignty*

Erik R. Seeman, *The Huron-Wendat Feast of the Dead: Indian-European Encounters in Early North America*

Peter Charles Hoffer, *When Benjamin Franklin Met the Reverend Whitefield: Enlightenment, Revival, and the Power of the Printed Word*

William Thomas Allison, *My Lai: An American Atrocity in the Vietnam War*

Peter Charles Hoffer, *Prelude to Revolution: The Salem Gunpowder Raid of 1775*

blood on steel

CHICAGO STEELWORKERS &
THE STRIKE OF 1937

MICHAEL DENNIS

Johns Hopkins University Press | *Baltimore*

© 2014 Johns Hopkins University Press
All rights reserved. Published 2014
Printed in the United States of America on acid-free paper
9 8 7 6 5 4 3 2

Johns Hopkins University Press
2715 North Charles Street
Baltimore, Maryland 21218-4363
www.press.jhu.edu

Library of Congress Cataloging-in-Publication Data

Dennis, Michael.
  Blood on steel : Chicago steelworkers and the strike of 1937 / Michael Dennis.
    p. cm. —(Witness to history)
  Includes bibliographical references and index.
  ISBN-13: 978-1-4214-1017-3 (hardcover : alk. paper)
  ISBN-13: 978-1-4214-1018-0 (pbk. : alk. paper)
  ISBN-13: 978-1-4214-1314-3 (electronic)
  ISBN-10: 1-4214-1017-6 (hardcover : alk. paper)
  ISBN-10: 1-4214-1018-4 (pbk. : alk. paper)
  ISBN-10: 1-4214-1314-0 (electronic)
    1. Memorial Day Massacre, Chicago, Ill., 1937. 2. Strikes and lockouts—Steel industry—Illinois—Chicago—History—20th century. 3. Steel industry and trade—Illinois—Chicago—Employees—History—20th century. 4. Industrial relations—Illinois—Chicago—History—20th century. I. Title.
  HD5325.I32U625 2014
  331.892'8691420977311090449—dc23   2013043618

A catalog record for this book is available from the British Library.

Special discounts are available for bulk purchases of this book. For more information, please contact Special Sales at specialsales@jh.edu.

Johns Hopkins University Press uses environmentally friendly book materials, including recycled text paper that is composed of at least 30 percent post-consumer waste, whenever possible.

CONTENTS

	Preface	vii
	Acknowledgments	ix
	Prologue: The Making of a Memorial	1
1	From Crisis to Confrontation	7
2	The Rising Tide of Rebellion	24
3	Memorial Day 1937	36
4	Red Scare and Popular Resistance	62
5	Little Steel and Class Warfare	94
	Epilogue: Rethinking the Massacre	110
	Notes	119
	Suggested Further Reading	131
	Index	135

PREFACE

In a decade in which violent labor struggles often dominated the headlines, the shooting and killing of 10 marchers and the wounding of more than 100 others still captured national attention. It was an episode that shocked even seasoned labor activists. By 1937, Americans routinely read about how policemen beat, gassed, and in some cases shot striking workers. But this event was different, and not just because of the appalling death toll. It captured national attention because—behind the police line, on top of a truck—a Paramount newsreel camera whirred. Journalists also captured disturbing images of the clash. While eyewitness accounts were maddeningly divergent, the film offered incontrovertible proof that the police had indeed assaulted unarmed American citizens. Of the mortally wounded that day, seven had been shot in the back. No less troubling was the celluloid evidence of burly Chicago police officers relentlessly pursuing and beating terrified marchers.

This is the least contested part of the story, however. The more challenging question is, what did the event mean in the context of the social upheaval of the 1930s? For too long, historians have treated the episode either as a dramatic but minor moment in a larger union struggle or as a particularly glaring example of the police brutality that characterized the struggles of the era. Yet the Memorial Day Massacre crystallized the most fundamental issues raised by the Great Depression. More than this, it took place at a particularly volatile moment in the movement for the rights of labor: 1937. That year, a massive wave of sit-down strikes broke out across the United States as workers took it upon themselves to achieve the promises of the New Deal. Business across the country reeled as workers engaged in this bold tactic for winning higher wages and gaining dignity in the workplace. When the police opened fire on the steelworkers that day, they also took aim at the social movement that challenged the authority of management to dictate the terms and conditions under which average Americans earned a living.

As a note before we begin, the present work—particularly, but not exclusively, chapters 4 and 5—draws in part from my book, *The Memorial Day Massacre and the Movement for Industrial Democracy*, published in 2010. It is reproduced with permission of Palgrave Macmillan.

ACKNOWLEDGMENTS

For his commitment to this project, his confidence in the author, and his critical acumen and wisdom along the way, I'd like to express my sincere gratitude to senior editor Bob Brugger. Equally invaluable has been the contribution of his assistant, Melissa Solarz, whose organizational and administrative skills, not to mention unfailing patience, kept this project on track when treacherous terrain threatened to derail it. I'm grateful to the anonymous reviewer as well, who raised questions that needed raising while pointing to areas that required additional exploring. With terrific skill and historical sensitivity, Kathleen Capels copyedited the manuscript; I am enormously grateful for her expertise, and so too, reader, should you be. I also owe a debt of gratitude to Kimberly Johnson, who provided indispensable supervision at the critical page proof stage—thank you. Finally, to Rod Sellers of the Southeast Historical Society, for his enduring commitment to telling the story of the steelworkers and their struggle for justice in modern America, my admiration and gratitude.

Blood on Steel

Prologue
The Making of a Memorial

ON A WARM SUNDAY in May 1937, as Chicagoans flooded onto the Lake Michigan beaches and streamed into Wisconsin for a weekend at the lakes, a group in the southeastern part of the city held a picnic. An unusually large group, totaling more than 2,000 men, women, and children, had gathered at Sam's Place on the Southeast Side, which had become an unofficial strike headquarters for the steelworkers who had walked out on the Republic Steel Corporation's Burley Avenue plant only the week before. A festive and energetic mood pervaded the event, as children played tag and skipped rope while adults sought out the shade, enjoyed refreshments, and caught up on the latest news. On this Memorial Day weekend, summer felt imminent.

The gathering was no ordinary picnic, however. Middle-class supporters and working-class activists had organized a protest against a steel company determined to hold out against the drive for unionization in 1937. Republic Steel joined Youngstown Sheet and Tube, Bethlehem Steel, and a handful of other steel companies to resist the union effort that had only recently succeeded at US Steel, the nation's biggest steel manufacturer. These "Little Steel" companies, which were little only in relation to US Steel, virulently opposed the demands of the Steel Workers Organizing Committee (SWOC) for union recognition and collective bargaining. Corporate executives—convinced that they could not afford the cost of a union contract as their firms recovered from the Depression, determined not to give the labor movement a foothold, and vehemently contesting the idea that employees should be able to limit management's power in any fashion—planned to hold the antiunion line. Even though 85,000 workers had walked out of Little Steel mills stretching from Pennsylvania to Illinois, the executives calculated that they could win.

At Sam's Place on Chicago's Southeast Side, workers and others in favor of their cause gathered in search of moral support from each other. Policemen had already attacked the steelworkers and their allies when they tried to picket the plant's gates. Now the demonstrators needed to renew their commitment to a struggle that had sent workers to the hospital and to jail. The evidence of police harassment could be found in the bandages wrapped around several of the protestors' heads that had been struck by police billy clubs earlier that week. The atmosphere may have been festive, but an undercurrent of anxiety permeated the ranks of those who converged at Sam's Place that day. Even so, few could anticipate how deadly serious that afternoon's event would soon become.

Those who arrived on the Southeast Side to show their support had responded to a call for a mass meeting sent out by steel union officials in the aftermath of that week's police brutality. Labor singers and a left-wing theater troop provided entertainment and boosted morale, but the speeches took center stage. In these evocative addresses, labor activists Leo Kryczki and Joe Weber castigated the Chicago Police Department's recent violations of civil liberties and reminded the crowd that the Roosevelt administration and the Supreme Court stood firmly in support of labor's cause. The speakers called on those present to exercise the right to picket, which Mayor Edward Kelly had reaffirmed after the violent attacks by police the previous week. Most importantly, they elicited a resolution to march to the gates of Republic Steel's Burley Avenue mill to protest the police harassment. Yet the mass demonstration would also be directed against the nonstriking workers who remained inside the plant to continue production—or, at least, to maintain the illusion of production. Those who were still appearing to work mattered enormously. If even a minority of the employees stayed inside, it would support management's case that the steelworkers suffered debilitating divisions. If divided, they could be conquered.

The resolutions were passed, the speeches were made, and the picnickers then formed a makeshift column that headed across the barren field by Sam's Place toward Republic Steel. That column included women, children, and other supporters who had never set foot inside a steel mill. Like so many demonstrations in the tumultuous 1930s, this one was made up of many parts of the working-class community. At Sam's Place, they found camaraderie. There, they built solidarity—an old term for class unity that is all but lost to us, but would have carried tremendous meaning for those in the social movements of the 1930s. Marching to Republic Steel, the protestors would put that solidarity to the test. By the time the demonstration ended, Sam's Place would be transformed into a bloody and frantic triage unit. The attending doctor and nurse, both supporters of the labor movement, would report heads split open by baton blows, flesh torn by bullet rounds, and mangled limbs that had to be amputated. On Memorial Day 1937, the violence that marred the long struggle for industrial democracy erupted once again. This time, it would come in the form of a terrifying, one-sided police riot.

Yet the march to the Republic Steel mill did not simply belong to a long lineage of antilabor episodes in American history. Instead, it vividly symbolized the larger movement for industrial unionism and social democracy that occurred in the 1930s. In the most intense period of class conflict that the United States had yet seen, industrial laborers, working-class supporters, and their middle-class allies revitalized the movement for workers' rights that had dissipated in the 1920s. The campaign to end the autocratic authority of factory management stood at the center of the labor upheaval of the 1930s.[1] In place of despotism, workers imagined a system of democratic representation for industrial laborers in the mines and mills. They believed that through union organization, they would achieve material improvements that had been denied to them by profitable but tight-fisted corporations. Equally importantly, they would achieve a measure of control over the pace of production that would reduce the constant dangers of modern industrial work. That control would help them limit the companies' relentless drive to intensify production, a process that saw employees working more hours, at a faster speed, while being paid less and less. It would give workers a chance to experience the economic security that President Roosevelt championed; it would give them the opportunity to become more fully human, to enjoy the promise of individual freedom that is at the core of the celebrated American dream. Moreover, it would allow them a greater sense of dignity in the work they performed. That control would end what the laborers considered industrial serfdom. It would establish the principle that workers did not have to relinquish their basic American rights when they took their place on the factory floor.

This drive for what labor reformers called "industrial democracy" attracted the support of middle-class liberals who saw the labor movement as a struggle for human rights, as well as a vehicle for broader social reforms. Teachers, social workers, writers, civil servants, community activists, and others believed that the labor movement would advance the values of social responsibility over selfish individualism. By challenging the managerial authority of the steel executives in the mills, the movement would also contest the cultural and political authority of corporate leaders, bankers, and financiers in American society. Middle-class supporters of the labor movement attached their own aspirations for economic democracy, social security, and political accountability to the steel workers' movement. The push for justice in the steel industry became a key point of convergence for an interracial political alliance that coalesced around Roosevelt.

In the years of the Great Depression, these middle-class Americans could hardly feel secure in their own status. Unemployment, financial insecurity, and uncertainty about the future plagued them almost as much as it did industrial laborers. In 1935, 14,000 Chicago teachers demonstrated outside of banks and financial institutions in protest against the city's failure to pay them their salaries. Even though they had been tear gassed and attacked by Chicago police in 1935, many of these same teachers would show support for the steelworkers. About 20,000 high school students would also launch a sympathy strike in support of the teachers. Some of these would walk the picket lines at Republic Steel and attend the rally on Memorial Day.[2]

Middle-class activists certainly joined the steel movement out of a sincere desire to promote social justice, but they also did so for personal reasons. The unsettling experience of losing their grip on middle-class stability had encouraged them to forge a larger coalition for social change. As labor writer and strike observer Mary Heaton Vorse explained, the devastating dislocations of the Great Depression had "swept away the old firm moorings of middle-class life. A mounting sense of insecurity had invaded millions of once independent and self-reliant lives."[3] She noted that many middle-class Americans, previously rooted in occupational mobility and property ownership, found their predicament during the Great Depression almost "indistinguishable from that of labor today." For a brief period, this shared sense of insecurity and of hope for enduring change produced a powerful alliance for reform.

This interracial alliance pointed the way toward a more egalitarian America. When the boisterous column of Memorial Day demonstrators marched off toward the Republic Steel plant in 1937, it included Mexican steelworker Max Guzman and Mexican immigrant Guadalupe Marshall, a social worker and activist based at Jane Addams's famous Hull House. African Americans Hank Johnson and William Young participated, as did Scottish immigrant and skilled steelworker George Patterson. SWOC's Women's Auxiliary decisively contributed to the Little Steel Strike, which explains why George Patterson's wife Dorothy joined the scene that formed the backdrop for the march to the Republic Steel mill. While Dorothy tended to the kitchen at Sam's Place, George joined the marchers, eventually becoming its de facto leader. Yet the interracial, cross-class alliance that surged to the front of the Southeast Side protest that day would quickly find that the group's every move had been tracked by local authorities. In addition to the police informants who infiltrated the steelworkers' ranks that day, uniformed Police Captain James Mooney and Officer George Higgins arrived on the scene. Both had been tasked with handling the labor protests of the early 1930s. Their experiences that day would eclipse anything they had previously encountered at an unemployed rally or a march in defense of Republican Spain as it struggled against the fascist uprising that threatened to topple the democratically-elected government.

It is essential to understand, then, that this shocking episode of antilabor violence did not represent an aberration in an otherwise peaceful history of labor relations. From the Great Railroad Strike of 1877 through to the Haymarket Incident of 1886 and on to the garment workers' uprising in 1909, people who took to the streets to test what they described as "wage slavery" could expect to confront bullets and batons. Similarly, authorities could expect to find men and women more than willing to eject scabs from a plant, fight police in the streets, and launch mass demonstrations designed to halt production. In contrast to the ineffective ritual of walking a picket line today, workers in the late nineteenth and early twentieth centuries engaged in militant protests that many of us would find shocking. Their actions certainly offer a challenge to the contemporary myth that American workers have always accepted the ethic of competitive individualism. In those days, workers and their families often gave as good as they got in picket-line altercations. They understood the issues at stake in a strike. They knew that the authorities had lined up against them, and that winning required them to impede production. In a sense, then, the Memorial Day episode can be seen as one point on a historical continuum of antilabor violence and trade-union militancy.

In another sense, however, the entirely unique features of the demonstration stand out, because on this day, Paramount newsreels caught the scene. For the first time, the media would provide a celluloid record of the antilabor aggression that had confronted workers trying to organize at least since the Gilded Age that followed the Civil War. The film vindicated the demonstrators' claim that Chicago police officers fired into and then pursued and beat fleeing demonstrators. It captured the men in blue treating wounded marchers with callous disregard. It also validated what labor activists had been saying for decades: in showdowns between workers and the police, the fight was rarely fair. The latter usually had superior firepower and generally inflicted the greatest damage. Although the newsreel footage would never be able to answer who or what provoked the clash between the marchers and the authorities, it left many conscientious people wondering if the virus of fascism had not reached American shores.

Even so, in this fractious era of labor unrest, the shooting and killing of 10 marchers and the vicious wounding of more than 100 others constituted merely the most extreme episode of coercion by the state. "The use of the police by the mills to shoot steel workers asking for their constitutional rights," Vorse observed, "is an old story. The shooting of workers in steel began in Homestead in 1892 and has gone on ever since."[4] Equally important, the sight of hundreds of workers engaging in a mass demonstration did not deviate from the norm in the 1930s. What really stands out about the Memorial Day incident is how it became the focal point for a larger movement—to revitalize the idea of American equality.

1 From Crisis to Confrontation

THE CAMPAIGN TO ORGANIZE the steelworkers formed the overarching context for the Memorial Day Massacre. The Little Steel Strike became a titanic clash between business and labor, but it was not the biggest strike of the decade. Nor, for that matter, was it a success for the labor movement. Then why was it important? For the Congress of Industrial Organizations (CIO), which split from the American Federation of Labor (AFL) in 1935, a victory in the steel industry promised strategic and symbolic gains. Winning here would accelerate the movement to organize semiskilled workers, who constituted the majority of industrial laborers in America. Since steel was critical to the economy, organizing the employees that produced it was equally crucial. A breakthrough would also overturn the reputation that the steel industry had forged as a bastion of antiunionism. The industry won that reputation when Andrew Carnegie defeated efforts by the Amalgamated Association of Iron and Steel Workers in 1894; it consolidated it in 1919, when US Steel utterly demolished the AFL's effort to organize the unskilled workers. Gaining a union contract in 1937 would prove that Goliath *could* be beaten after all.

In order to grasp the significance of the Memorial Day Massacre, we need to understand that those who supported the steel strike believed that it spearheaded a challenge to an American form of dictatorship. The notoriously antiunion steel industry had come to symbolize the unilateral authority of business in American society. Despite the calamity of the Great Depression, the leaders of America's major industries clung to the prerogatives of managerial control. To the surging labor movement of the 1930s, steel magnates appeared to be the palace guard for the "economic royalists" whom Roosevelt and his supporters considered to be the enemies of New Deal reform.[1] As the La Follette Civil Liberties Committee observed in its 1939 report on private police systems, "company owned and controlled towns . . . have created conditions approximating industrial peonage. . . . A company town is an autocracy within a democracy. It has no law and no control save the interests of the company."[2] Workers went on strike not simply to achieve higher wages, but to oppose the industrial despotism under which they labored. That same industrial tyranny, which seemed to privilege profit margins over human needs, also appeared to be responsible for the economic crisis then gripping the nation.

The Little Steel Strike frames the events that occurred in Chicago's Southeast Side. More than an extension of the CIO's campaign to organize the steel industry, the Little Steel Strike produced a community uprising that briefly united working people and middle-class reformers. Together, they challenged the dominance of American business in social affairs. By this time, a coalition of urban middle-class reformers, progressive intellectuals, industrial laborers, northern African Americans, and southern whites had already coalesced under the umbrella of the Democratic Party. It was this alliance that reelected Roosevelt in 1936. The same coalition took action during the Little Steel Strike. The strike provided an opportunity for these groups to articulate their own vision of social democratic reform and realize some of its ideals. It created the environment in which a contest over union representation escalated into a fundamental examination of the character of American society. It was a pivotal moment in what Hungarian economist Karl Polanyi described as the modern effort to impose social protections where only business-friendly policies had prevailed.[3] Seen from this perspective, the steel strike in Chicago expressed the reformist energies that had developed in the early years of the Great Depression.

While SWOC's leaders focused on winning a contract, rank-and-file workers drew the connection between the drive for union representation and the movement for social democracy. Achieving democracy in the workplace meant opposing petty authoritarianism, political exclusion, and racial inequality. It also signified promoting a greater measure of economic justice and ethnic pluralism, while adopting a more cosmopolitan vision of what constituted an American identity.[4] Above all, it required them to confronting the threat of international fascism.[5] More than a struggle over a "piece of paper," the Little Steel Strike gave rise to a community revolt that vividly expressed the democratic impulse of the 1930s.

The Great Depression was great not only because of the magnitude of the economic calamity it produced, but also because it dealt a powerful blow to the myths that had governed American society for so long. Historian Ian McKay has described it as a "structure-shifting 'matrix event'" that produces a "sudden and drastic moment of rebellion" in which large collections of people simply reject the social structures imposed on them by addressing the immediate source of oppression and challenging the larger structure that sustains it.[6] They begin to move toward "supersedure," a moment when people start to understand that the current injustices they are experiencing have emerged from contradictions in the greater social and economic system. The period from late 1936 to the summer of 1937 was a pivotal moment *within* the larger crisis of the Depression. The goals of SWOC's leaders may have been limited, but the Little Steel Strike precipitated what was surely another matrix event, when workers from across the nation began the wave of sit-down strikes that directly challenged the claim that property rights gave business the authority to do what it wanted to its workers in the factories and fields of America. By taking over factories, mines, and retail outlets across the country, tens of thousands of laborers protested against their employers' determination to ignore the new federal law that protected workers' rights on the job.

More than a struggle for union recognition, the Little Steel Strike expressed the egalitarian themes of the era. It also produced a more progressive vision of labor unionism than the leaders of the steel union had in mind. Ultimately, the importance of the Little Steel Strike resides in its challenge to the ruthless competition and social inequality inherited from the Gilded Age. In doing so, it became part of a larger reorientation of American values in the era of the Great Depression.

Labor and Roosevelt's New Deal

The national exasperation at Herbert Hoover's timid handling of the economic crisis during the Depression paved the way for the election of Franklin D. Roosevelt, governor of New York, to the presidency of the United States in 1932. Americans voted against what seemed like Hoover's inaction, even though Hoover was certainly not indifferent to the plight of the unemployed. What he could not do was wrench himself and the United States out of a tradition that dictated limited federal government activism in the face of an economic crisis. In 1932, it seemed that Franklin D. Roosevelt could. So Americans voted against Hoover, but they also voted for Roosevelt's jaunty optimism, his bold approach to the crisis as governor of New York, and his pragmatic willingness to question some of the cherished economic dogmas of the era.[7]

Yet it was his support for the humane values of the Progressive Era and his determination to achieve a measure of economic security for average Americans that defined his presidency. It became the glue holding together the political coalition that Roosevelt built during his first term. That alliance included liberal reformers, middle-class social activists, African American northerners, and white southerners, with urban industrial workers forming its linchpin. The labor upheaval that led to the events in Chicago propelled this coalition into action. Yet evidence of a multiracial working-class movement on the march would increasingly alienate the white segregationist southerners who provided vital support for Roosevelt's legislative agenda. The political expediency of including the Jim Crow south in the Democratic coalition would soon become a major liability to progressive reform.

Even before Roosevelt's historic election in 1932, American workers started moving from shock to indignation over the calamity of the Great Crash. They rediscovered the possibilities of collective action, mass demonstrations, and industrial unionism. The conditions that produced widespread despair also fostered the regeneration of social protest.[8] In the desolate early years of the Great Depression, the Communist Party emerged as a decisive force in social protest. Inspired by the Russian Revolution, it coalesced in the years of labor upheaval and government repression of left-wing activists that followed America's intervention in the war. In the 1920s the party grew, but it remained a radical faction spouting revolutionary doctrine that most American workers found alienating and irrelevant. By 1931, the Chicago branch could only count 2,000 dues-paying members in its ranks. Yet it was able to attract around 12,000 votes for William Z. Foster, the Communist Party's candidate for the presidency in 1932 and veteran of the steel strike of 1919. The crisis of the Great Depression provided an opportunity for those who dreamed of a new socialist day.

Determined to organize the dispossessed, the party aggressively recruited African Americans. By defending the Scottsboro Boys, nine black teenagers unjustly accused of raping two white women on a train to Alabama in 1931, the party earned the respect of many black Chicagoans. It advanced its reputation for promoting racial equality when it nominated a black candidate for the vice presidency in 1932. Communist Party activists further solidified their credibility as defenders of the disadvantaged by organizing the unemployed, sponsoring mass demonstrations against racial discrimination on Chicago's South Side, advocating for evicted black tenants, and even moving the evictees' belongings back into their homes after they had been tossed onto the street by building managers and municipal authorities. However dubious its allegiance to the Soviet Union, however strident its rhetoric, and however deluded its notion about the possibility of revolution in the United States, this was still an organization that acted on its convictions.[9] That determination to put their bodies on the line and wade into the most threatening of situations in order to win a strike or stop an eviction inspired many who would never think of joining the party, but were more than willing to follow those who could provide decisive leadership.

Defending evicted tenants was part of a larger national strategy to represent the legions of unemployed desperate for at least some relief. Confronting insufficient state and federal efforts to ameliorate the crisis, the party set out to unite communist trade union activists and the unemployed in a movement for relief. Starting in Detroit, these councils spread to Buffalo, New York; New York City; Philadelphia; Cleveland; Los Angeles; and Chicago. Planning an International Unemployment Day demonstration in 1930, party activists got a taste of what the Chicago police force had in store for radicals. Echoing the antilabor violence of incidents such as the Pullman Strike of 1894, Chicago police raided a meeting of the Unemployment Day organizers at Mechanics Hall. They arrested 13 leaders on charges of sedition and hauled them off to the downtown station. There, police subjected them to "the third degree," which included exhausting interrogation and physical beatings. Despite the systematic intimidation, the Communist Party activists carried on and successfully staged the March 6 unemployed demonstration. In all, hundreds of thousands of Americans joined that protest. Workers in the United States would no longer accept their predicament.[10]

While authorities moved to crush this incipient class revolt, the party adopted a stronger focus on immediate issues. Party leaders realized that revolutionary sloganeering could not attract workers to a social movement. Average Americans worried about where their next meal was coming from, not about the theory of capitalism's imminent collapse and the possibilities of world revolution. In July 1930, the party formed the Unemployed Councils of the USA. At a conference in Chicago, the delegates endorsed the policies of federal relief, unemployment insurance, democratic control of relief expenditures, and racial unity on behalf of the Depression's victims. Communist organizer Steve Nelson understood the transformative significance of the party's commitment to the Unemployed Councils. "It was from involvement in the daily struggles that we learned to shift away from a narrow, dogmatic approach. . . . We began to raise demands for immediate relief by the city and state, immediate assistance to the unemployed, and a moratorium on mortgages, and finally we began to talk about the need for national unemployment insurance."[11] Hunger marches on state capitals and on Washington, DC; rent strikes against predatory landlords; protests at relief stations; demonstrations on behalf of the unemployed; and rallies against racial discrimination raised the profile of the party. They also sent the message that those left destitute by the Great Depression would not quietly accept their fate. Rather than simply being a sycophantic champion of Soviet foreign policy, the Unemployment Councils constituted a grassroots organization dedicated to racial equality and industrial democracy.[12]

With the election of Roosevelt and the revival of social protest, a widening band of average Americans began to question the system that had produced so much misery. Urban housewives, destitute farmers, unemployed industrial laborers, and bankrupt shop owners began to believe that government had a responsibility to assist the most disadvantaged. Together with left-wing activists and liberal supporters, they made the case that it was incumbent on governments to alleviate unemployment and the social distress of the Great Depression. But they also sought to rejuvenate the industrial union movement. Rejecting the racist exclusivity of the AFL craft unions, they organized on the basis of interracial cooperation. This democratic activism provided a catalyst for the labor movement of the 1930s. Out of the Unemployed Councils emerged men and women trained in the techniques of community organization and political leadership. They would utilize those skills in the drive to organize steelworkers.[13]

New Deal policy also proved crucial to the revival of the labor movement. Roosevelt's National Industrial Recovery Act of 1933 finally gave industrial workers a foundation for claiming that labor unionism was legitimate and necessary. Through Section 7(a) of the act, workers gained the right to form unions of their own choosing and to enter into collective bargaining with their employers. It also provided for the maximum allowable number of working hours and, for some industries, minimum wages. To American manufacturing employees, who were ravaged by the Great Depression and accustomed to instability, it seemed as if a new day and a new deal had really arrived.

Yet government planners did not simply act out of a sense of benevolence for industrial employees. The idea behind the National Industrial Recovery Act was to reduce and control industrial production, thus raising prices and liquidating inventories. The other part of the equation was to increase consumers' buying power and counter the tendency toward excessive concentration of economic power in the hands of too few corporations. In the ideal scenario, consumer representatives and labor activists would balance the profit-making impulse of big business by protecting decent wages and reasonable prices. Through labor organization and consumer activism, the Roosevelt administration hoped to restore buying power and foster business recovery.

Few corporate leaders agreed. Large companies thwarted the efforts to control production, frequently ignored the appeals by consumers and small businesses for price controls, and consistently sought ways to undermine Section 7(a). When the Supreme Court struck down the National Industrial Recovery Act, which was an experiment in national economic planning, in 1935, it also neutralized the hope that the federal government might provide an umbrella of protection to stave off the torrent of antilabor tactics that companies deployed during the Great Depression.

Even so, Section 7(a) represented an enormous breakthrough for organized labor. Section 7(a)—together with the Norris-LaGuardia Act of 1932, which prohibited the hated court injunctions that companies had used since the Gilded Age to prevent picketing and break strikes—powerfully stimulated the labor movement and challenged the idea that economic recovery would simply restore business as usual. These two pieces of legislation also set the precedent for government involvement in a labor market that was often chaotic, volatile, and unfair. "When progressive Democrats convinced a reluctant Roosevelt to replace the now defunct Section 7(a) with the National Labor Relations Act, they validated the experiment in collective bargaining, which had electrified the ranks of organized labor and stimulated the movement's recovery. This new act encapsulated the desire for a more humane but also stable system of labor-management relations. The National Labor Relations Act (also known as the Wagner Act for its lead sponsor, New York senator Robert L. Wagner, a veteran of the Progressive Era who could still remember the futile efforts of garment workers to organize a union before the terrible Triangle Shirtwaist Factory Fire of 1911) provided a government-protected mechanism for collective bargaining. In addition, the bill safeguarded the principle of democratic workplace representation and prohibited a battery of antiunion measures that businesses had employed time and again to undermine the labor movement. The Wagner Act was a pivotal breakthrough for industrial unionism. This act was not simply a labor bill, but part of a larger social policy aimed at protecting the poor, redistributing income, and limiting the power of the corporations that had dominated American society since the Gilded Age. It was Wagner's hope that the measure would do more than create a mechanism for collective bargaining; he imagined that it would promote sensible cooperation between workers and their employers and foster the idea of industrial democracy that the steelworkers now championed.[14]

Yet the Wagner Act was also designed to ensure industrial peace by limiting the opportunities for workers to protest company policies. Although it protected collective bargaining, it did so by appealing to the economic expediency of interstate commerce, not to the constitutional principles of freedom of assembly and freedom of speech. Instead of promoting the act as a defense of basic human rights, Wagner and the Roosevelt administration justified the measure as a means of ensuring continuous production. Rather than touting it as an instrument for achieving justice in the workplace, the proponents of the Wagner Act argued that it would promote economic recovery by reducing the maldistribution of wealth that had supposedly triggered the Great Depression.[15] Basing the measure on economic efficiency rather than on constitutional principles, however, gave opponents a massive political hammer to use against strikes. A strike was the one weapon that workers had successfully used to challenge managerial dominance, but instead of defending their right to throw down their tools and walk out, the framers of the bill emphasized how collective bargaining would minimize the need for this confrontational tactic. Whatever claims labor unionists might make about the issues involved in a given strike, opponents could argue that it disrupted commerce, infringed on the supposed rights of private property, and violated the intent behind the protection of collective bargaining. Because of this statutory foundation, each time there was a conflict between the workers' right of assembly and the property rights of the business owners, the employers could argue that the right to property was guaranteed by the constitution, while the right to organize a union was only protected in so far as it promoted commerce. And if going on strike impeded commerce, why not do everything possible to prevent strikes in the first place?[16]

Over time, the federal government and the Supreme Court would prohibit the tactics that workers had used in their uneven fight against dictatorial managers. Most disastrously, companies would be permitted to hire replacement workers during a strike, which would severely restrict the ability of employees to do the one thing they had to in order to win concessions: inflict economic pain on their employers by stopping production. Much of this would develop in the postwar era, but it was already clear by 1935 that federal government protection came at a price. That included the rising influence of union bureaucrats, the growing importance of contractual legalism, and the diminishing ability of workers on the shop floor to oppose the authority of management. At the time, however, most industrial workers cheered the arrival of the Wagner Act.

Most also understood that industrial unionism did not come to American workplaces simply because the federal government willed it. The growing disaffection of workers on the front line and their increasing determination to challenge the company unions propelled sympathetic legislators to act. In the dynamic that often develops in moments of social upheaval, grassroots discontent pushed lawmakers to accede to at least some of their demands. If revolution was not imminent, widespread social disorder certainly was. Even though the Communists and the Socialists had gained only a small proportion of voters, continued indifference toward the plight of the "forgotten man" could lead voters to endorse more radical alternatives to the Democrats and the Republicans. Labor activism and social protest pressured political leaders to adopt the hallmark features of the New Deal state. It was certainly in Roosevelt's interests to do so, since so many of these second-generation ethnic Americans represented potential Democratic votes in 1936. It was also to the benefit of industrial capitalism, since festering discontent could produce the kind of instability that might derail the whole free enterprise system once and for all.

A New Kind of Unionism

This new mood of labor confidence produced a revolt within the ranks of the union movement that would reverberate from Washington, DC, to Chicago and beyond. In November 1935, only months after the Wagner Act became law, dissident members of the American Federation of Labor formed the Committee of Industrial Organizations. Eventually ejected by the AFL in 1938, the latter would become the Congress of Industrial Organizations. Led by John L. Lewis of the United Mine Workers, the CIO rejected the political conservatism and craft exclusiveness of the AFL. Instead, Lewis, labor leader Sidney Hillman, and their AFL allies championed industrial unionism, irrespective of employees' skill levels. That meant that the CIO would organize all the workers, not just the skilled artisans, in a given enterprise, such as the steel, automotive, electrical, or meatpacking industry. The idea was to standardize wages and prevent companies from exploiting their workforce in order to earn higher profits. Equally important, the CIO would organize workers across racial, ethnic, and gender lines. By advocating industrial unionism and espousing an egalitarian philosophy that echoed the populist themes of the New Deal, the CIO helped crystallize a mass movement.

At the same time, labor activists experimented with new techniques that would change the balance of power inside the factory. In January 1936, rubber workers in Akron, Ohio, went on strike without ever leaving the plant. In a coordinated action, they shut off the machines to protest a wage reduction and the firing of union members. Instead of walking out of the plant, they sat down at the machines in which they believed they had invested sweat equity. What they had retrieved from labor's heritage was one of the most powerful tools available to those who try to challenge the dominance of management: the sit-down strike. In Flint, Michigan, autoworkers deployed the sit-down technique so successfully that General Motors was finally compelled to recognize their union. That effort initiated a wave of strikes that would mark 1937 as one of the most rebellious years in American history. Electrical employees, furniture workers, leatherworkers, oil workers, retail clerks, municipal employees, and steelworkers soon used this tactic, because it gave them control over a production process that often seemed completely autocratic. It was an effective technique because it minimized violence, prevented the company from introducing strikebreakers, and imposed a choke hold at the key location: the point of production. Sit-downs allowed workers to take direct control over their own affairs.[17]

By 1937, the wave of sit-down strikes had rolled into Chicago. "Now the auto workers were in the height of their 'sit-downs' and the steel workers was [sic] right there with them," steelworker George Patterson remembered. "Small plants on the west side of Chicago began organizing by sitting down on the job, and then calling our office for card applications."[18] The sit-down strikes won gains for workers that had eluded them in the many years of corporate welfare and company unionism. Employee sit-downs broke out in Chicago at Valley Manufacturing, Chicago Steel and Wire, the Wilson and Bennet Company, and the Burnside Steel Foundry. From Chicago Heights to Aurora, De Kalb, and Waukegan, SWOC organizers scrambled to stay ahead of the spreading wildfire of sit-down strikes. While district subdirector Nick Fontecchio tried to get a handle on the Chicago-Calumet region, Meyer Adelman organized in Wisconsin, where industrial unionism caught on with equal fury. The most brazen sit-down strike took place at the Fansteel Metallurgical Corporation in North Chicago. There, workers battled police in the streets. Policemen then constructed a scaffold to assault the sit-down strikers. Using tear gas, they routed the workers lodged on the second and third floors of the factory. Class warfare had come to Main Street USA.

In this period of intensifying class conflict, no industry figured more prominently as a battleground than steel. Yet one of the most important breakthroughs for labor came when US Steel agreed to recognize the Steel Workers Organizing Committee. The CIO had created this organization for the express purpose of bringing unionism to the steel industry. Under the leadership of John L. Lewis, the United Mine Workers spearheaded SWOC, supplying many of its key organizers and most of its money.

Lewis and Philip Murray, who became president of SWOC, also recruited Communist Party activists—who had proven to be tough, skillful, and implacable organizers—to conduct field work. These party activists provided vital leadership, fostering a sense of class awareness and political coherence among workers. They articulated the employees' grievances and channeled them into effective protests.[19] Communist Party activists who worked for SWOC collaborated with the party's members in Local 1033 (the fledgling union outpost at Republic Steel). African American labor activists Frank Edwards, Ben Patterson, and Henderson Daniels belonged to the party and worked as committed organizers for the steel union and as advocates for social justice. The communist-influenced International Workers Order would prove to be invaluable as it generated support for SWOC among foreign-born and second-generation Eastern European workers.

More than any other labor confrontation, the sit-down strike at General Motors' Flint, Michigan, plant decisively shaped the situation in the steel industry. In January 1937, CIO president John L. Lewis negotiated a settlement with US Steel president Myron Taylor that allowed SWOC to become the bargaining agent for its members. Taylor would not accept a union shop allowing SWOC to have exclusive bargaining rights, but the agreement was a watershed accomplishment for the labor movement. For the first time, a major steel manufacturer had accepted the legitimacy of industrial unionism. US Steel acceded to the 8-hour day, the 40-hour week, and the principle of time and a half for overtime. In addition, SWOC won a 10-cent wage increase that brought the workers' hourly rate to 62.5 cents an hour. That had been the objective since the beginning of the organizing drive. Since the contract also included provisions for seniority and a grievance procedure, union negotiators could be confident that they had won a substantial victory.[20]

Why had the notoriously antiunion US Steel Corporation capitulated to SWOC? Taylor had to consider the costs and consequences of a Flint-type strike happening at his company's own Carnegie-Illinois steel plant. "The General Motors strike showed United States Steel how tightly an industry can be closed down with this new-fangled industrial unionism," observed *Nation* magazine, "and after it was all over the union won anyway."[21] Labor writer Art Preis has observed that Lewis could not be held singularly responsible for the agreement. Instead, Lewis had "the able aid of a 'negotiating committee' of 140,000 GM [General Motors] sit-downers, particularly the brave auto workers of Flint who held the GM plants for 44 days."[22] Union bargaining succeeded because US Steel could afford the pay increase. By 1936, the company had returned to profitability, earning $67 million in 1936 and $130 million in 1937. Taylor understood that a protracted fight against unionization would cost the company dearly.[23] The SWOC contract at US Steel ultimately depended on the organization of rank-and-file workers.[24]

The Taylor-Lewis settlement decisively accelerated the budding labor movement. Within the steel district in Illinois and Indiana, union lodges in towns from Chicago Heights to Gary signed a manifesto that expressed their members' new mood of determination. They believed that collective bargaining would prevent a return to "the tragic days of the past and make the future, not only of the steel workers but of the entire nation, secure."[25] The US Steel contract accelerated the union movement among those in the industrial labor force. Around 35,000 workers signed SWOC cards in the first two weeks following the Taylor-Lewis deal. By April 1, 1937, 59 companies had signed contracts with SWOC. More than 200,000 steelworkers now belonged to the new union.[26]

Although liberal reformers celebrated the settlement, it met with a cool response from Republic Steel president Tom Girdler and his Little Steel allies. It "came not only as a surprise," labor journalist Mary Heaton Vorse reported, "but as a betrayal to the heads of the big independents in steel, the irreconcilable enemies of union labor."[27] Tom Girdler became the lightning rod for that sentiment among the chiefs of Little Steel. "Unquestionably many thousands of workmen interpreted this event as a wonderful victory for themselves," he noted bitterly in his autobiography, "but I am sure that thousands of workers were shocked, even horrified by the news."[28] Little Steel would not sign an agreement, Girdler believed, "because we were convinced that a surrender to the C.I.O. was a bad thing for our companies, for our employees; indeed for the United States of America. A majority of our employees did not belong to C.I.O. and we were not going to force them in against their wishes." Instead, Girdler defiantly remembered, "we were determined to fight."

At least at that moment, it seemed as if Girdler and Little Steel had chosen the wrong side. On April 12, 1937, in the *National Labor Relations Board v. Jones & Laughlin Steel Corporation* decision, the Supreme Court upheld the Wagner Act; the company was now required to rehire several employees that it had dismissed for union activity. Despite the opposition of the industry's American Iron and Steel Institute and the National Association of Manufacturers, the Supreme Court had prohibited the antilabor practices that had been the mainstay of steel manufacturing.[29] It seemed as if steel unionism had gained an unstoppable momentum. In Aliquippa, Pennsylvania, a company town that Girdler had fashioned into a "Little Siberia," SWOC launched a strike that galvanized union support. When the workers at the Jones and Laughlin steel plant walked out on May 12, they signed SWOC cards by the hundreds. In two days, the company had capitulated. It would also sign the watershed agreement already reached between US Steel and SWOC. The steel union won a National Labor Relations Board vote by a landslide. All signs suggested that Little Steel was about to capitulate.[30]

The *Jones & Laughlin* decision, however, did not signal an unqualified victory for labor. While the Supreme Court upheld the Wagner Act, it also stated that the law *did not require* the company to sign a contract that officially recognized the union.[31] That clause steeled Girdler's opposition to the union. Castigating the CIO as an allegedly racketeering, communist-dominated outfit, he dismissed SWOC's demand for a settlement. On the other side of the contest, however, the US Steel contract and the *Jones & Laughlin* agreement convinced SWOC to take a hard line. On May 3, SWOC's Clinton Golden sent a telegram to Republic Steel executives, claiming that the company's intransigence had produced "widespread unrest among employees." Golden announced that if the managers didn't sign a contract, "we shall be obliged to disavow all responsibility [for] union members in your mills remaining at work."[32]

Girdler had no interest in conciliating Golden and the steel unionists. In fact, Girdler eagerly anticipated a confrontation. Ironically enough, he had the Wagner Act on his side. He justified his defiance by appealing to the loophole in the Wagner Act, which required companies to bargain, but not to sign.[33] The conflict at Little Steel now became a contest for union recognition. It would test whether the union could establish a foundation of legitimacy in the steel industry. Philip Murray, John L. Lewis, and SWOC regional director Van Bittner believed that only through a signed contract could workers introduce something like the rule of law into the steel mills. Republic Steel bolstered its position by granting its employees the same wage increase that workers in US Steel's Carnegie-Illinois plant had won in the Taylor-Lewis agreement. Now the main point of contention between workers and Republic Steel was union recognition. Girdler remained determined not to concede it.[34] Moreover, just as the SWOC campaign intensified, the company unleashed an all-out assault on union supporters in each of its mills. As the National Labor Relations Board observed, Republic Steel launched a "campaign of intimidation and coercion by spreading propaganda in favor of the [Employee Representation] Plan and against the C.I.O. at meetings of the employees called by them for that purpose."[35] The media ignored these aggressive antiunion efforts, as would the opinion pollsters. According to company propaganda, its executives did not want to disrupt the "present harmonious relations" that supposedly prevailed at Republic Steel.[36] Meanwhile, SWOC support steadily grew.

The Widening Social Movement

It is easy enough to conclude from all of this that the labor struggles of the 1930s featured hard-bitten union leaders squaring off against equally cunning business moguls. But that would misrepresent the larger forces at play. The most profound changes developed at the shop-floor level and in the streets of Chicago's Southeast Side. There, workers who had submitted to an arbitrary and unaccountable management for far too long finally gained the courage to oppose it. Many of these workers had voted for Roosevelt, received relief from the federal government, or worked at a job provided by the New Deal's Public Works Administration or the Works Progress Administration. Most had experienced a period of unemployment. As companies slashed corporate welfare programs, workers came to understand the benefits of the federal government's intervention. At the same time, industrial laborers knew that real change in the workplace depended on them. When they joined the CIO movement, they participated in a political transformation that had started during the Unemployed Council protests of the early 1930s. They began to imagine themselves as agents of their own destiny.

More than this, they began to act according to a set of values that challenged the dominant ideas of the steel owners and coal operators. The strange alchemy that occurred in dozens of clashes during that tumultuous decade now materialized during the Little Steel Strike. The limited issues that supposedly defined the controversy morphed into something larger and more profound. From a dispute over wages or union recognition, the labor struggles of the 1930s frequently escalated into battles over basic civil rights, democratic citizenship, and social democracy. Provoked by the police and thwarted in their efforts to fulfill the promises of the New Deal, strike after strike became an occasion for widening the sense of class consciousness that defined America in the 1930s.

The democratic and pluralistic ideals of the New Deal era also appealed to working-class women. Soon after the formation of an independent union at US Steel's Carnegie-Illinois South Works plant, Dorothy Patterson joined a handful of others in launching the Women's Auxiliary. Long before the steelworkers of Chicago hit the pavement, these women mobilized resources, undertook clerical work, and galvanized support for the union. The South Works auxiliary also sponsored a Women's Steel Conference to push the movement ahead. "The central auxiliary would knock on every door in Chicago Heights," Mrs. Patterson declared, "to show the wives of steel workers the value of unions."[37] The auxiliary also utilized its connections with trade unions, fraternal orders, civic groups, women's trade unions, and social clubs. "Every day brings thousands of workers into the union. You can be instrumental in bringing the message of organization to other thousands, through the women in your own trade union or club."[38] Auxiliary members could thrive in a movement that allowed women a greater measure of personal and political autonomy than most had ever experienced.

Primarily designed to win male support for the union, the women's auxiliaries nevertheless advanced a more ambitious vision of social change. On the one hand, they operated within the boundaries of union authority and occupied a subordinate role that reflected organized labor's tendency to portray the movement as an exclusively male affair. On the other hand, the egalitarian ethos of the period gave women the opportunity to define their own concept of social justice unionism.[39] "We are fighting mad about our homeless youth and all these miserable, unattached women, some sleeping in hallways, in railroad stations and wandering around in boxcars. We demand a better life, or even, as the Spanish women, the right to defend our homes against these great corporations."[40] The Great Depression convinced many working-class women that the home could not be insulated from the world of social and political conflict. In effect, they wanted the domestic sphere to become a check on the ruthless acquisitiveness of modern America. Women's auxiliaries now began to embrace ideas that opposed the dogma of unrestricted competition. They promoted the view that labor issues had a direct impact on society, and that women had a stake in the public forum. This vision of community unionism developed into a central theme of what was to become the Popular Front movement.[41]

The coming conflict could not be limited merely to union recognition. It quickly developed into a contest to determine if the cooperative ideals of the Roosevelt era would prevail. "Huge corporations, such as United States Steel and General Motors, have a moral and public responsibility," John Lewis announced to a national radio audience in the midst of the Flint sit-down strike. "They have neither the moral nor the legal right to rule as autocrats over the hundreds of thousands of employees. They have no right to transgress the law which gives to the worker the right of self-organization and collective bargaining. They have no right in a political democracy to withhold the rights of a free people." Lewis did not acknowledge, however, that the struggle for "the rights of a free people" also transpired within the union.[42]

As the labor movement revived, it drew middle-class Americans—progressives, intellectuals, and social activists—into an alliance. These groups began to link their hopes for economic security to an interracial labor movement, forming what was described as a "popular front." For years, critics dismissed the Popular Front movement in the United States as a creation of the American Communist Party, working on behalf of the Soviet Union. More recently, historians have reexamined these conclusions, finding in the movement something much more dynamic, egalitarian, and reformist. The Communist Party USA profoundly influenced the movement, but it never dominated it. Instead, the Popular Front participated in a larger coalition of forces rooted in the CIO, aligned with the policies of the New Deal, and committed to the principles of racial equality, industrial unionism, and antifascism, both at home and abroad.[43]

By summer 1937, Republic Steel workers who supported SWOC wanted more than the 40-hour work week; they wanted to eradicate the grinding sense of indignity that had afflicted work in the industry since US Steel crushed the strike of 1919. Now, a signed contract became inseparable from the drive to overturn industrial despotism. Yet for Republic Steel's management, SWOC's insistence on a contract amounted to sheer temerity. Steel industry executives like Tom Girdler considered SWOC and its CIO directors to be subversives who did not respect the basic American prerogatives of property ownership and managerial control. Since they had conceded the US Steel wage increase, they argued that the demand for union recognition was an abstract issue at best. The ground for compromise buckled beneath them both.

2 The Rising Tide of Rebellion

BY MAY 1937, union support in southeastern Chicago's steel district had reached a crescendo. Gus Yuratovac, Emil Badornac, Joe Germano, and a cadre of SWOC organizers had steadily dissolved the ingrained culture of antiunionism at Republic Steel. Workers at the small, independent companies of Calumet Steel and Inland Steel in Chicago Heights had organized thoroughly and anticipated action. At Youngstown Sheet and Tube in Indiana Harbor (a neighborhood in East Chicago, Indiana, just over the state line from Chicago), 3,821 workers out of a total of about 5,500 had signed SWOC cards.[1] "Republic Steel ready for the break," Germano reported on May 13. Four days later, he observed "marvelous enthusiasm from the 500 present" at a SWOC meeting of Republic workers. By May 20, John Riffe noted that the "meetings are better than they have been." By May 24, Germano was reporting that the "Republic Steel situation [is] becoming serious due to large number of layoff[s]. Men demanding strike. Everything under control."[2] Frontline workers now dictated the pace of events.

While SWOC organized, Republic Steel stocked weapons and armaments for an imminent confrontation. The company had spent more than the city of Chicago on tear gas and sickening, or vomiting, gas. Its inventory included 4 submachine guns, 525 revolvers, 64 rifles, 245 shotguns, and enough clubs and ammunition to hold off the Illinois National Guard.[3] Republic's contingent of 370 police guards stood ready to use them. The company's escalating layoffs, harassment, and antiunion espionage aggravated an already drum-tight tension.

The struggle was not only in the steel mills and the local community, but within the area of the law. In Illinois, state law protected legal strikes and placed no prohibition on mass picketing. The Chicago Department of Law had assured the city and the public of those protections no more than two months before the Little Steel Strike. In an opinion bulletin dated March 31, 1937, Chicago's Corporation Counsel, Barnet Hodes, declared that the police department should not interfere with picketing when it was "conducted in a peaceable manner, for picketing as the term is generally understood is lawful in Illinois when peaceable and without coercion, intimidation, or disorder."[4] The memo stated that "relief" should be sought from the courts, not the police, a reminder that the "thin blue line" was not the final arbiter of the law. Referring to the state's anti-injunction act, Hodes pointed out that "workers should not be interfered with by the government authorities in the conduct of lawful strikes in a lawful manner." In his position as Corporation Counsel, Hodes maintained that picketing *was* a legitimate tactic in a strike. Moreover, both Illinois and federal law supported it.

Perhaps most surprisingly, the Chicago Department of Law admitted the legality of the fundamental question involved in a strike. "The strike is a lawful instrument in a lawful economic struggle or competition between employer and employees." Hodes may not have had mass picketing in mind when he drafted this opinion. Still, it illustrates how the labor upheaval of the 1930s was altering inherited ideas about law. Workers did not wait for the police or the courts to interpret the meaning of the law—they would advance their own. Yet they would also challenge the Corporation Counsel's version of a legitimate protest. Without applying enough pressure on the workers remaining inside the plant to shut down production, a strike would be ineffective.[5]

Police harassment was now a constant companion in the campaign for workers' rights. The behavior of the police force toward labor activists in the period leading up to the Little Steel Strike made it clear that such actions would continue, despite the opinions of Chicago's Department of Law. On

April 9, around 200 workers at the Walker Vehicle Company on West 87th Street went on strike. Captain James L. Mooney, who would soon gain notoriety in the Memorial Day Massacre, quickly dispatched 300 constables to the vicinity. According to SWOC lawyer Paul Glaser, policemen "roughly handled" the strikers.[6] They "dispersed, shoved around and struck the pickets and strikers," attacking them as far as five and six blocks away from the company's premises, and finally limiting the demonstrators to two inconsequential pickets. A month later, the police broke up a strike at the Crowe Name Plate Company on North Ravenswood Avenue. Glaser described the "police terror" that prevented pickets or strikers from getting close to the factory: "Strikers were arrested, stopped and shoved and beaten blocks away from the plant."[7] Captain Lynch had command of the forces in the area; when Glaser tried to remind him of the Corporation Counsel's opinion, Lynch pointed out the lay of the land: Barnet Hodes was not in charge of the Chicago Police Department.

Serving and Protecting the Few: The Police in Chicago

Since the police department was a key agent in the events surrounding Memorial Day 1937, it's important to understand the attitudes that shaped their response to the labor movement. The officers who confronted the crowd at Republic Steel carried a set of assumptions influenced by their experience of marches by the unemployed and by relief-station demonstrations in the early 1930s. In those previous actions, the police force confronted militant participants who took direct action in order to get their message across. These mass demonstrations tested the nerve and discipline of even the best officers. The protestors, often galvanized into action by Communist Party activists, frequently fought back against police provocation. Conducting sit-ins at relief stations and noisy marches into Chicago's Loop district, in the heart of the city, workers traumatized by the crisis of the Great Depression showed their determination to force the hand of change.

Chicago's police force found the marching masses alarming, but it was the ethnic and racial composition of the demonstrators that disturbed them most. Eastern European industrial workers joined African Americans, Hispanics, and native-stock whites in protests that seemed to threaten a new working-class coalition. This alignment conjured up images from the city's troubled history of poisoned race relations and class conflict. The notion that ethnic minorities produced subversive troublemakers was a legacy of the steel strike of 1919. At that time, US Steel and its media allies asserted that the strike was the work of Bolshevik agitators, not a workforce groaning under the weight of 12-hour days and laboring in appalling conditions for paltry wages. Those responsible for defending the city's political and economic order in the 1930s had absorbed the lesson from the 1919 strike: coercion worked, particularly when it pitted whites against blacks and ethnic minorities. Even though this attitude ignored the racist practices and industrial exploitation that produced these events, it governed the perspective of policemen on the front lines in the 1930s. The presence of radical activists in the Unemployed Councils only reinforced long-standing stereotypes about the link between Eastern European immigrants and labor agitation. Seeing a racially diverse, ethnically complex working class willing to take to the streets under the leadership of known radicals seemed to confirm the worst suspicions of the police. Quite simply, it allowed them to minimize the genuine grievances that motivated Chicagoans to take collective action while scapegoating the devious "Reds."

The racial and class antagonisms of 1919 shaped the subculture of policing in the 1930s, but the assumptions that informed it developed out of the labor strife of the Gilded Age. It was then that Chicago's upper class came to the conclusion that labor activism was an occasion for seditious and violent behavior. The Great Railroad Strike of 1877, in which striking railroad workers and their supporters shut down the nation's transportation system in protest against drastic wage cuts, demonstrated the complicity between the Chicago police and Big Business. In pitched street battles, the police killed as many as 35 workers and helped quell the most explosive working-class rebellion of the Gilded Age. In 1885, local authorities mobilized the police force to suppress a streetcar strike that had paralyzed the city. Literally hundreds of workers suffered blows from the policemen's billy clubs as constables protected scab streetcar operators in downtown Chicago.

Following the Haymarket Bombing of 1886, the police carried out the nation's first major Red Scare. City elites sent the message that the violation of civil liberties and the disruption of peaceful labor unionism was a small price to pay in order to round up the alleged anarchist conspirators. During the Pullman Strike of 1894, Chicago's finest would provide the frontline defense of industry's property rights. In the stockyards and in the Teamsters' strikes of 1904 and 1905, city authorities, operating on behalf of industrial leaders, sent a phalanx of police officers to beat strikers into submission and ensure safe passage for replacement workers.

The police force and company officials often worked in close cooperation. In some cases, officers held on to suspected ringleaders so company thugs could then work them over. When literally tens of thousands of female garment workers went on strike in 1910, policemen became the shock troops for the textile industry. They arrested demonstrators on the flimsiest of pretexts and beat those who resisted. Police headquarters even dispatched officers to specific factories, where they would operate under the supervision of textile management, functioning as both strikebreakers and the defenders of the prerogatives of private property. By the beginning of the twentieth century, a close alliance had developed between the police force and the political establishment. In labor disputes, the men in blue now functioned primarily as strikebreakers, not impartial defenders of the law.

The violence that the police inflicted on labor activists was a natural extension of the belief that coercion was necessary to maintain order. The labor upheaval of 1919 seemed to reinforce that nineteenth-century assumption. Yet violence wasn't just reserved for labor agitators; it was a routine feature of policing in a city known for lawlessness, political corruption, and social instability. Attacking striking workers differed little from clobbering juvenile delinquents; according to the police, both groups defied public order. Moreover, police violence in defense of public order had the support of Chicago's middle class. In the early twentieth century, constables frequently arrested young, male, homeless drifters because they represented a drain on municipal resources and a threat to social stability. Even so, police officers routinely provided shelter for the homeless during the winter months. They would arrest vagrants and provide them with a meager form of social assistance. In the rough-and-tumble reality of urban America in the Gilded Age, the police operated as both thuggish enforcers *and* ministering angels.

Despite the growing professionalism of the police force, it continued to operate on a set of values that privileged community stability over objective legal standards. In part this resulted from the fact that well into the twentieth century, recruits received little formal training. By 1929, they spent four weeks at the police academy before hitting the streets. Approximately one-fourth of that time was devoted to parade square drill, and another quarter on laws and city ordinances. Supervisors expected rookies to memorize the list of crimes in alphabetical order. The final test must not have been very demanding, though, since it was reported that no recruit had ever failed.

Once on the force, police officers became intimately involved in the neighborhoods in which they patrolled. Policemen were poorly supervised, and many spent their time in the barbershops and saloons on their beat. Departmental reforms in the twentieth century eventually tightened control over beat cops, but the relationship between the frontline officer and the neighborhood remained strong. It is little wonder, then, that by the 1930s, vagrancy and disorderly conduct constituted between 40 and 66 percent of all police arrests. Constables believed that it was their principal duty to protect the community against drifters, tramps, and troublemaking labor agitators. The police often made little distinction between them. And while the force itself was as ethnically diverse as the city it patrolled, 76 percent of the police captains in 1930 were of Irish descent. Police officers of Irish and German extraction dominated the force by the time the Little Steel Strike broke out. Law enforcement offered an avenue of social mobility for skilled and semiskilled workers, although they never led comfortable lives. Moreover, frontline police officers did not belong to the city's elite—they came exclusively from the ranks of the working class. Even so, at least they did not have to labor in the packinghouses or the steel mills. The men who donned the police force's blue uniform prized their tenuous status.

In a city marred by ethnic and racial tension, many who had "made it" had contempt for those beneath them. Police officers considered the semiskilled African American, Croatian, Serbian, Italian, and Jewish workers to be their social inferiors. That latent animosity flared when workers walked the picket lines. Even before the labor upheavals of the 1930s, however, police routinely targeted Mexicans for arbitrary arrest and raided Mexican commercial establishments. In one incident, the police arrested 20 Mexicans for the "offense" of standing outside during a dance at Hull House. For the police, Hull House came to represent left-wing activism, immigrant subversion, and ethnic differences. The Chicago police had been operating on the same racial prejudices at least since the vicious race riot of 1919, when understaffing on the force and public support for rioting whites created the conditions in which 23 blacks and 15 whites would be killed, and hundreds of others injured. On Memorial Day 1937, Guadalupe Marshall, a Hull House social worker active in the Mexican expatriate community, would discover how deeply those prejudices continued to run in the police force.

The police department that Chicago's CIO activists confronted was also highly politicized. Departmental appointments came down a chain of political influence that stretched from city hall to the local precinct. Mayors selected police chiefs from the ranks of police captains, while ward politicians supervised the appointment of the police brass who served in their districts. Examining Chicago's political machine in the 1930s, *Fortune* magazine concluded that around some "50,000 jobholders" had won their positions through political patronage. Court bailiffs, health department employees, sanitation workers, board of education appointees, secretaries, city hall scrubwomen, firefighters, and roughly 7,000 police officers owed their positions to city hall. Advancement through the police hierarchy required officers to appease the political establishment. Police brass expected officers to perform political services for the ward organization if they wanted to keep their jobs. Selling tickets for events that generated cash for the local ward heeler was as important as issuing tickets for traffic violations.

Equally important, the police had to know which gamblers, two-bit hustlers, and brothel madams had protection from the local ward. If they harassed those who had obtained the favor of the local establishment, they could find themselves assigned to "the woods," a precinct nowhere near their own neighborhood.[8] Civic reformers who wanted to clean up and professionalize the police department found it appalling that criminal syndicates paid for protection. The city's bootlegging, prostitution, gambling, and illicit saloon operations often had connections with local politicians. Criminals further down the pipeline, some as lowly as pickpockets and burglars, also had patrons in the establishment. It wasn't just the petty criminals who had ties to the elite. The ward bosses controlled the patronage that kept the city running. Graft and corruption oiled the wheels of the entire machine. As *Fortune* magazine noted in 1936, even Chicago's underclass depended on the political establishment. "There are landladies who wish [for] lax inspection from the fire department and health department. There are employees and cousins and uncles and aunts of all these people. There are dope peddlers, beggars, and pickpockets. All of these people need protection."[9] By the 1930s, the police subculture fostered during the turbulent Gilded Age was still abundantly evident.

If anything, the antiradical, xenophobic element within the police department had intensified. When Chicago police lieutenant Make Mills appeared in 1930 before the Fish Committee, a forerunner of the anticommunist investigations of the 1940s, he made it clear that the department would not allow constitutional civil liberties to get in the way of its defense of the status quo. As commander of the Radical Squad, the predecessor of the Red Squad, Mills conducted surveillance on left-wing demonstrations and public meetings. He also revealed some of the tactics his squad used to disrupt Unemployed Council demonstrations. In some cases, the police simply denied permits for the marchers; in others, they delayed issuing the permits in order to make the protestors susceptible to arrest. Mills and his minions also helped break strikes by providing dossiers of information to the antilabor newspapers. Labor activists and unemployed organizers recalled how the Radical Squad would drive patrol cars into assembled crowds, impose arbitrary time limits on political speeches, and physically attack unemployed demonstrators. Make Mills and his men offered living proof that the spirit of the Haymarket hysteria persisted into the 1930s.

Prelude to Memorial Day

The momentum that had been building culminated when the Steel Workers Organizing Committee called a strike on May 26, 1937. Girdler had already locked out the workers at Republic Steel's Massillon mill in Ohio on May 20. In response, SWOC called for the complete immobilization of Republic Steel and Youngstown Sheet and Tube. Republic Steel executives remained defiant: they would continue to operate at the company's mills in Buffalo, New York, and in Canton, Warren, and Niles, Ohio. Most fatefully, they decided to continue production in Chicago's Southeast Side.[10] The Little Steel Strike would soon affect seven states and 90,000 steelworkers. In the early stages of the conflict, however, Chicago would be its epicenter.

The steelworkers in the Southeast Side now champed at the bit. At 3:00 p.m. on Wednesday, May 26, around 200 employees converged at the gates of the Republic Steel plant on Burley Avenue, well in advance of the union's scheduled walkout time. Captain James Mooney and his superior, John C. Prendergast, who was chief of the police force's uniformed branch, desperately wanted to prevent union supporters from launching a sit-down strike. Mooney claimed that as many as 1,500 workers had remained inside the plant, 200 of whom had stopped working and "would not go out." Soon after Mooney arrived at the steel mill, SWOC subdirector John Riffe approached him and asked to be allowed to address the workers inside. Instead of consulting those workers who had engaged in a sit-down strike, Riffe sought to appease the captain. "I talked to him and I told him our people who are on strike here is [sic] going to conduct this according to the law."[11] Riffe and SWOC officials wanted to avoid "trouble." Determined to appear respectable, SWOC's leaders wanted to dampen the militant energy that had broken out in the steel district.

Captain Mooney wanted the SWOC subdirector's assistance in "cleaning out" the workers on the inside, and Riffe was prepared to give it. Escorted by plainclothes police officers past roughly 150 to 200 uniformed policemen who had been housed and fed at Republic Steel's expense, Riffe entered the steel mill. There, he found 200 to 300 employees engaged in the tactic that had made the Akron, Ohio, and Flint, Michigan, strikes famous. Riffe "asked them if they knew me and they said 'yes.'" He then explained the deal he had struck at the very beginning of the strike. "Well, the management and the captain [Mooney] have requested me to ask you men to all leave this plant, they don't want any sit-down strike and we don't want any of our men violating the law." But there was no law prohibiting sit-down strikes, at least not yet. A seasoned labor activist, Riffe was nevertheless conflating "the law" with the will of the Chicago Police Department. He appealed to the sit-down strikers to leave the plant and join the picketing outside.[12]

Riffe then exited the steel mill, but he pleaded with Mooney to establish a strike protocol. "As long as you stay across the railroad you can picket," Mooney announced. "You won't be bothered." Even in that ineffective position, striking workers would find themselves harassed by the police. Returning to the picketing strikers on Burley Avenue, Riffe looked back to see anywhere between 300 and 400 workers streaming out of Republic Steel. Those employees could have become the nucleus of an effective sit-down campaign. Yet Riffe had made it clear that SWOC would not support a sit-down strike, the most potent weapon in organized labor's arsenal in the 1930s. At a critical moment in the strike, Riffe squandered the opportunity to tip the balance in favor of the steelworkers by maintaining the presence of sit-down strikers inside the plant. On the outside, the police and Republic Steel's security force would have a clear advantage.

Even though SWOC's leaders strangled this effort at direct action inside the mill, it is important to understand how the events throughout the strike zone expressed the spirit of the sit-down wave of 1937. It was this moment, when workers experimented with oppositional tactics but also took actions that challenged their own leadership, which defined the Little Steel Strike at the ground level. John L. Lewis and Philip Murray did not simply dictate union strategy. Galvanized into action by the example of the sit-down strike and the promise of federal government protection, workers engaged in the kind of self-determination that defined the various movements for social justice in modern America. Pouring out of the Republic Steel mill, the workers cheered each other in a fashion that seemed to counter the accumulated indignities of years spent performing subservient work under arbitrary authority. From that altered perspective, illustrated by their unauthorized sit-down action, striking employees at Republic Steel's Burley Avenue plant created the possibility for a more dynamic movement for social change. Moreover, some of those steelworkers had no interest in relinquishing the sit-down tactic.[13]

Almost as soon as SWOC organizers had established a functioning picket line, the police attacked.[14] When John Riffe returned to the mill, he discovered almost 150 police officers marching out of the plant's gates in ranks of three. Closing quickly on the motley picket line, the policemen clubbed the unarmed marchers and ordered them back. Riffe urged the workers to hold the line while he appealed to the authorities to respect their right to demonstrate. Captain Mooney and another uniformed officer took hold of Riffe and dragged him to a patrol wagon. Powerless to intervene, Riffe watched the police reprise the part they had played in countless earlier strikes. "They kept beating them back, pushing them back, hitting at them with their sticks, breaking the picket line. . . . They [the picketers] was broke [sic]."[15] By that time, about 300 police officers were patrolling the Republic Steel plant. Others began to pursue and attack the retreating strikers. SWOC organizer George Patterson recalled that the police "went up on porches, into the taverns, and into people's gangways between the homes, on to the property, driving everyone away from the Republic gates." The police systematically repelled workers at the plant gates, striking anyone who resisted. They had "cleaned the street and no one was allowed on the porches," Patterson testified.[16] "No one was allowed anywhere."

Repulsed by the police, a group of strikers now desperately launched a spontaneous sit-in. They sat down "in several inches of mud in defiance of being moved back from the gates," the *Chicago Herald and Examiner* reported.[17] While he was in retreat, Patterson recalled seeing "a bunch of the men sitting there, so I sat down along with them."[18] The authorities wasted no time in disrupting this demonstration. "I never saw so many people getting picked up and hurled bodily into the patrol car," Patterson recounted. Police threw Patterson and Women's Auxiliary member Marjorie Minors into the paddy wagon. Lacking any kind of coordination, the sit-in was no match against the police attack. Yet the workers' determination to utilize the tactics of direct action, improvise on the front lines, assert their claim to citizenship in public spaces, and challenge the presumption of authorities to dictate the law would continue throughout the entire strike.

Contrary to their role as law enforcement officers, the police had violated labor law. In fact, the very next day Mayor Edward S. Kelly produced a statement reaffirming the steelworkers' right to peaceful picketing. Union president Gus Yuratovac was determined to test Kelly's guarantee. Following the altercation of Wednesday evening, Captain Thomas Kilroy briefly permitted a picket line of eight strikers on Thursday, May 27. The officers on the ground soon disrupted it, however, sending the workers back across "the prairie," an open stretch of land between the railroad tracks that ran alongside the steel mill and Green Bay Avenue to the east.[19] Encouraged by Mayor Kelly's announcement, Yuratovac formed a line of strikers and set out for the plant. They advanced no farther than 117th Street and Green Bay Avenue. There, police intercepted them. Told they could not picket the gate at the steel mill, the marchers produced a copy of the *Chicago Daily Times*, with Mayor Kelly's statement prominently featured. Yuratovac later testified that the ranking officer "used profane language toward the paper and he asked if Mayor Kelly was higher that they was."[20] The marchers presented no real threat to the bluecoats. In effect, the police had *become* the law in the Southeast Side. The closest they would permit the pickets to get to the plant was three blocks away.

In a situation in which the Burley Avenue Republic Steel plant continued to operate, token picketing would have no impact on the outcome of the strike. Strike leaders would have to find a way to apply decisive pressure on the company. That was particularly urgent, since as many as 300 nonstriking workers remained inside. The police finally permitted six pickets in front of Republic's gates. Yet the police stationed at the steel mill insisted that these six maintain a circular pattern on Burley Avenue, ordering them not to stop at any point and prohibiting them from handing out literature or speaking with any of the nonstriking workers. SWOC district subdirector Nick Fontecchio then decided that it was time to escalate the pressure. In the electrified atmosphere that accompanies any confrontation, both the leaders and the rank and file struggled to respond to a quickly shifting situation. Riffe had capitulated to the police; Fontecchio was prepared for direct action.

On Thursday evening, one night after police had beaten and harassed picketing workers for exercising their rights under the Wagner Act, SWOC led a column of around 700 strikers and supporters down Green Bay Avenue in the direction of Republic Steel. Carrying a massive American flag, the group advanced as far as 117th Street and Burley Avenue. There, police officers formed a defensive line. As soon as one of the officers struck Dominick Esposito, the flag bearer, the police waded into the crowds, clubs swinging.[21] The march degenerated into a wild melee. Steelworker and SWOC organizer George Patterson observed "people being clubbed all around me." He detected the unmistakable odor of gunpowder as the police fired two or three shots into the air. Policemen swung at the strikers, many of whom responded in kind. Some of the workers hurled rocks at the police, while others used picket signs and fists against officers wielding hickory billy clubs and .38s. Striker Ben Mitckess was knocked unconscious; Lucille Koch, a particularly committed member of the Women's Auxiliary, also fell to the ground under the policemen's blows. The constables soon had the disoriented marchers in retreat. Pushed back to 117th Street and Green Bay Avenue, the strikers stopped, facing a solid line of bluecoats "swinging their clubs back and forth."[22] In all, the police wounded 18 strikers, 3 seriously enough to require hospitalization.

Shocked by the lawmen's aggressive response, the marchers now confronted the silent cordon of police officers. One woman had retrieved the group's American flag and advanced to the front of the assembly, along with Patterson and a few others. "This is what you have done to us in the fight we had," Patterson exclaimed as he pointed to one of the wounded strikers, whose head was bleeding profusely. The steelworkers took other injured marchers, some "bleeding very badly, their heads . . . split wide open," and one apparently knocked unconscious, forward to the front of the line. Patterson continued: "We only asked for our right to go through and picket peacefully. You should not have done this." Demoralized and stunned by the attack, the marchers sullenly returned to Sam's Place.[23] Somehow they would have to establish a mass picket line.

3 Memorial Day 1937

IN SUMMER 1937, the movement for democracy in the steel industry became the focus of a wider campaign to transform American culture itself. The Memorial Day demonstration on May 30 against police brutality toward Republic Steel workers who picketed on May 26 and 27 became the fulcrum of the Popular Front movement in Chicago, and the Little Steel Strike was its lightning rod. Popular Front groups such as the Chicago Repertory Group set the tone. In 1936, they entertained unemployed workers and labor unions. By 1937, they had established a reputation as left-wing theater rebels. They performed Clifford Odette's *Waiting for Lefty*, Albert Maltz's *Black Pit*, the Federal Theater Project's *One Third of a Nation*, and other plays that reflected the themes of industrial democracy and antifascism. In 1938 they would perform an adaptation of Marc Blitzstein's *The Cradle Will Rock*. On Memorial Day 1937, they sang union songs to boost the morale of the battered rank-and-file workers. Standing on the back of a flatbed truck, student activist Mollie West sang "The Ballad of Joe Hill," "Solidarity Forever," and a selection of labor songs.

The Little Steel struggle also drew support from Chicago's middle-class progressives. Students and teachers from the Francis Parker School in North Chicago traveled to the gathering at Sam's Place on the Southeast Side. So, too, did theology students from the University of Chicago. Frank McCulloch, a member of the Council for Social Action of the Congregational Churches of America, attended the Memorial Day meeting, as did Congregational minister Chester B. Fisk, the son-in-law of Albert W. Palmer, president of the Chicago Theological Seminary (a South Side institution, adjacent to the University of Chicago). Fisk attended the meeting at the behest of Reverend Raymond Sanford, a Congregational minister who operated Common Ground, a liberal community action organization in the Southeast Side. Sanford would continue to support the workers during the Red Scare that followed the 1937 Memorial Day incident. Chicago writers Meyer Levin and Studs Terkel, the latter of whom was then a member of the Chicago Repertory Group, joined the demonstration. Dr. Lawrence Jacques, his wife, and three other physicians lent their services to the gathering that day. Jacques, sympathetic to the labor movement and a supporter of republican Spain, would play a decisive role that day. A collection of like-minded individuals—ministers, seminarians, teachers, social workers, students, journalists, labor activists, and civil libertarians—joined them. Women marched in strength on Memorial Day. According to some participants, about 300 women joined the rally and the march, and many of them brought their children along.[1]

Ultimately, more than 1,500 people, and as possibly as many as 2,000, attended the rally at Sam's Place, the unofficial strike headquarters only blocks away from the plant. After speeches decrying police injustice and the complicity of local officials, a majority voted to support a resolution to march to the plant and demonstrate for their right to picket. (As much as those sympathetic to the steelworkers have wanted to portray the resolution as a spontaneous outpouring of indignation, more than likely it was proposed at a union planning session the day before.) Critics of Communist Party involvement in the steel movement accused it of treachery when it was discovered that key activists such as Joe Weber and Hank Johnson did not join the march. Not only had communist organizers failed to participate, but union leaders did not show up as well. John Riffe was nowhere to be found. Nor, for that matter, was SWOC district subdirector Nick Fontecchio or regional director Van Bittner. The most crucial demonstration of united front solidarity would transpire without the leadership of union officials. It did not bode well for the movement. Alarmed at the apparent vacuum, SWOC organizer George Pat-

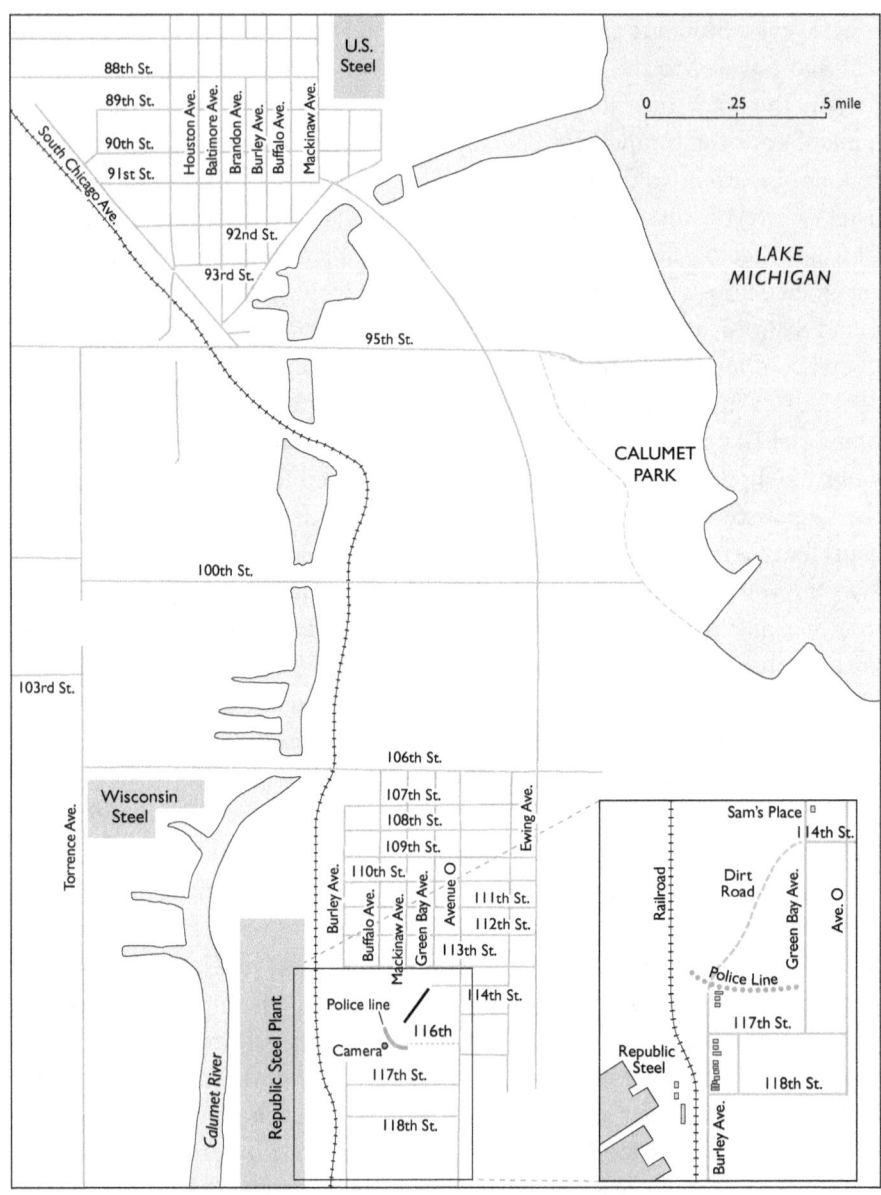

Chicago's Southeast Side. The inset shows the location of specific places along the route that marchers took from Sam's Place to the Republic Steel plant on May 30, 1937. Map by Bill Nelson.

terson took the lead. He was joined by Jim Stewart, a Scottish compatriot and president of SWOC Local 65. Patterson fell in at the front of the column, next to the flag bearers, as they moved steadily toward Burley Avenue and 117th Street. That was "the spot we always seemed to meet the police," Patterson recalled. He and the other marchers soon caught sight of a police cordon, a familiar—if threatening—presence in the continuing Republic Steel strike. The addition of several photographers and a Paramount newsreel truck to the scene was considerably less familiar. "It struck me as comical," Patterson recalled, "but I had little time to think. We were face to face with the police."[2]

In the twentieth-century era of film and radio, labor confrontations had become a public spectacle. The singing of spirited protest songs was a central part of that spectacle, as a combination of demonstrations and music was a signature feature of the labor movement of the 1930s. For a brief period, the egalitarian aspirations and emotional militancy of the CIO crusade decisively influenced popular culture. Mollie West, who was a high school student at the time of the Little Steel Strike, became a troubadour of the protest. She had initially become "interested and radicalized" by her involvement in a student strike. When high school authorities announced plans to cut extracurricular programs such as swimming, dancing, and music, the students began to organize a walkout, and West joined the strike committee. Working on picket signs at night, she and the other members of the executive committee could not believe their eyes when police raided the meeting and arrested the dissident students. Like so many labor activists of that era, an encounter with heavily arbitrary authority left an indelible impression on her.[3]

The convivial atmosphere after leaving Sam's Place had steadily deteriorated as marchers moved closer to the police line. Once face to face, the mood became increasingly tense. Guadalupe Marshall—who always went by "Lupe"—was a Hull House social worker and a labor activist strongly committed to her Mexican allies in the steel industry. Demonstrating that day, she remembered the "vile" language that an 18-year veteran officer, George Higgins, allegedly hurled at the female demonstrators. Higgins told a different story. According to him, the strikers threatened the police, supposedly shouting: "You lousy Chicago coppers, you or nobody else is going to stop us. We are going in that mill and drive them out." Captain Mooney declared that he was shocked by the incendiary language of the marchers. "I never saw such people in all the years of my experience," he observed. "They acted out of their heads, completely wild," using the "most profane words I ever heard.... The profanity was terrible." The officers also claimed that the marchers wielded more than inflammatory words. According to Officer Lawrence Lyons, almost every marcher within sight carried a club, a baseball bat, or a crowbar. Lyons and the others professed to have encountered an edgy crowd primed for a confrontation.[4]

The officers' accounts were diametrically opposed to the recollections of the marchers who confronted Captains Mooney and Kilroy, who both had been involved in the fracas at the Republic Steel plant the previous week. The writer and journalist Meyer Levin recalled that there were about 25 people nervously singing "Solidarity Forever," but their voices quickly tapered off as they approached the police. Once there, "the line was rather quiet." Flag bearer John Lotito was one of the few demonstrators who tried to negotiate with the police. He asked one of the young patrolmen why they would not simply allow the marchers to picket peacefully. "We ain't gonna cause no trouble," Lotito supposedly pleaded. "We are just going up to the corner there and show them people inside that we are on strike." Republic Steel employee Harry Harper was also at the front, vainly appealing to the police to allow him to search for his brother, Peter, who had remained inside when the strike began. "I looked at the officers' faces to my right and to my left. The faces were drawn, the lips quivering, and they seemed to be intoxicated with something that is hard for me to explain." Harper had the sense that "something was brewing," as did James Stewart, who also stood in the front line of the march. "Why don't you go back home?" one of the patrolmen apparently asked Stewart. "Why, we are only going to march past the plant and establish our right to peaceful picketing," he replied. Stewart was only partly disingenuous. The march was certainly "peaceful," but the demonstrators also wanted to put pressure on the strikebreaking workers camped out at Republic Steel, the only way in which they could hope to swing momentum toward the fledgling union.[5]

Then, without warning, the scene shifted. In the time it took for Orlando Lippert to change the lens in his Paramount newsreel camera, the stalemate had exploded in a flurry of gunfire. According to some eyewitnesses, the fusillade came immediately after Captain Mooney had read out loud a statement demanding that the marchers disperse. "As I was addressing one of the officers in front of me," Marshall remembered, "Mr. Higgins had moved away somewhat. The police were closing in, closing their ranks and crowding us, pushing us back all this time," compressing the marchers and aggravating the already tense situation. "There are enough of you men to march alongside of these people, to see that order is kept," Marshall anxiously announced to one officer. "Like hell!" he retorted. "Like hell! Like hell! Like hell there are!" Laughing "real sarcastically" at Marshall, the policeman muttered "something about sending these —— back." James Stewart recalled that right after Mooney's statement, a police officer lashed out at him. Another struck him in the forearm with his billy club. "From there on," Stewart remembered, "it was confusion."[6]

Frank McCulloch of the Council for Social Action had been observing the march from behind police lines. "It was the beginning of a backward movement," he observed, with "uncertainty, apparently, in the ranks of the marchers." McCulloch must have seen the pushing incident that Lupe Marshall noted. So, too, did Reverend Fisk, standing 50 yards behind the front line and carrying a handheld movie camera. Spotting a platoon marching to reinforce the police on the left flank, Fisk temporarily abandoned his camera and watched the scene unfold. "Immediately after that, there was a backward motion at the head of the crowd. I could not tell what caused it at all but could tell immediately that the people were giving way at the very head." This was the point when the blow from a billy club struck James Stewart. At the same time, Harry Harper, who had been looking for his nonstriking brother who had remained inside the plant, thought he heard "a blast of a whistle, and then all hell seemed to have let loose." Harper was the only marcher who noted the sound. Observing the standoff from the row of houses behind the police line, Frank McCulloch "saw a rock come from the ranks of the strikers and hit the fence at the end of the alley." Almost immediately he noticed clouds of tear gas rising above the prairie. It was at that moment, McCulloch believed, that the police launched their attack. Once the rock struck the fence, he saw one of the officers take out his weapon and fire it in the direction of the marchers, who were in full flight. "His gun was pointed directly in the direction of the men and not up in the air."[7] Others in the ranks of the sweating and nervous bluecoats quickly followed suit.

No sooner had the whistle been blown—or the column of marchers been pushed backward, or the rock hit the fence—than Harry Harper was struck on the left side of his head. He dropped like a sack of potatoes, "the blood . . . gushing out of my face. It was running in my mouth. I went back and held my hand above my eye. I was in a crouching position so the blood would not strangulate me. . . . I tried to retreat and go back, but . . . I had no vision in my left eye." Flag bearer John Lotito had been talking to one of the police officers when the bluecoats attacked, unexpectedly but furiously. "I got clubbed and I went down, and my flag fell down," Lotito remembered. When he tried to retrieve it and stand up, the police struck him again. "I was like a top, you know, spinning. I was dizzy. So I put my hand to my head, and there was blood all over. I started to crawl away. . . . I didn't know what I was doing."[8]

Without warning, the police had torn into the demonstrators' ranks. George Patterson had been one of the strikers appealing to the police to escort the marchers to the plant. After Captain Mooney finished reading aloud the statement calling on the marchers to disperse, the Scottish labor organizer once again tried to appeal to the officers to let the protestors establish a peaceful picket line. "He looked through me," Patterson noted, "down at me. Like he never saw nor heard me." As the police waded into the strikers, Patterson turned and ran. Lupe Marshall recalled that she was "still talking to these officers in front of me when I heard a dull thud toward the back of the—of my group." Mollie West, who only moments before had been singing "Solidarity Forever," now noticed an ominous cloud swirling in the grass ahead of her. It took only a moment for her to realize it was tear gas. "A lot of the people at the front began to wheeze and weren't able to breathe," she noted, and they began to move back.[9]

Tear gas canisters fell rapidly behind the front rank of marchers, but the pop of what many believed to be firecrackers or blanks soon riveted the crowd's attention. "When I was about fifty yards from the front of the crowd," theology student Marilee Kone recalled, "there was a sudden quick series of explosions, and a great mass of tear gas rose up at the front." At the same time that the tear gas canisters thudded to the ground on the prairie, another terrifying sound ripped the air. "Almost instantaneously," Kone noted, "there was a volley of shots." One crack was followed by a cluster of shots, which were soon consumed in a torrent of small-arms fire that engulfed the marchers in the center of the line. Reverend Fisk described simultaneous "explosions" breaking out at the front of the line, but he was not sure if they were tear gas or gunfire. At exactly the same moment, he noticed "a shower of stones and other missiles" descending out of the air, hurled from the back of the column of protestors. Police officers now fired directly into the mass of marchers. Some thought it was machine-gun fire. Marshall believed that the gunfire sounded "more like thunder." What had appeared to be almost a

"The Memorial Day Incident," depicting the first moments of the police assault. After a volley of gunfire was directed at the front ranks of the marchers, the police returned their revolvers to their holsters and took out the nonregulation truncheons provided to them by Republic Steel. The police would use these to devastating effect against the marchers, who are clearly in retreat. Note the white tear gas clouds, the tangle of marchers at the center of the photo, and the photographer at the bottom of the image, partially hidden by the right side of the tree. Records of the US Senate, ca. 1944, National Archives.

staged event only a minute before now dissolved into wild terror and confusion, with tear gas and flying billy clubs and whirring bullets seeming to erupt simultaneously. For several seconds, the barrage continued. Theology student J. Gordon Bennett, who had run ahead of his University of Chicago compatriots to a position approximately 70 yards from the center of the front line, heard a "burst of gunfire" explode at almost precisely the same moment the crowd turned and ran. "A bullet whizzed past. I dropped into a depression, then got up and continued running." Not far behind him, two marchers helped another who had been shot in the leg. A woman, also shot in the leg, hobbled to a nearby car; others carried a wounded 10-year-old boy to the vehicle.[10] The floodgate that had maintained a minimum of civility following the previous week's altercations had burst open.

The marchers had little sense of what had precipitated the assault, but they soon had no doubt about the impact of the revolver fire on their ranks. Despite his wounds, Harry Harper could see strikers "going down, as though being mowed down by a scythe." Looking to his right, he saw officers with drawn revolvers firing directly into the crowd. Harper immediately bolted for a ditch. Overcome by "the instinct of self-preservation," he ran, terrified and convinced that he would get a bullet in the back. As he dove for cover, Harper heard a plaintive voice calling out "Help me buddy, I'm shot," but he was in no condition to be able to do so. Hugging the ground, Harper was alarmed to see "a green ball of fire" fall to his right, only a few inches from his face, "spitting blue smoke" and irritating his injured eye. He tried to retreat in the direction of Sam's Place, but as soon as he stood up, "a terrible trembling feeling came over me and a sickening feeling in the pit of my stomach." Harper was lucky; grabbed under the arms by a marcher, he was hauled back to one of the cars that dashed to the grassy area once the shooting broke out.[11] Others would remain on the prairie.

Lupe Marshall, who had marched close to the front of the column, now found herself in the eye of the storm. Turning to look behind her, she saw that "the people that were standing in back of me were all lying on the ground, face down. I saw some splotches of blood on some of the fellows' shirts." Marchers piled one on top of another, some severely wounded, others knocked to the ground in the terrified retreat. Despite her diminutive size, the 4-foot 11-inch, 97-pound Marshall was not spared. Police officers struck her down on the spot, leaving her suffering from deep lacerations and bruises. Dazed and disoriented, she stood and tried to move around the fallen strikers. Marshall then spotted a police officer clubbing another marcher. As the fellow tried to raise himself up, the officer rained blow after blow on him. Grabbing one of his feet, the patrolman then turned the marcher over and began dragging him on his back. It was then that Marshall noticed that the striker's shirt was bloodstained. "Don't do that!" she screamed at the officer. "Can't you see he is terribly injured?" As soon as she uttered this plea, another officer struck her in the back and she fell to the dusty ground. A third patrolman delivered a punishing kick to her side. The tiny social worker tried to get to her feet, only to have yet another officer strike her three times on the back, dropping her to the ground again.[12]

The police hauled her unceremoniously to her feet and dragged her to a patrol wagon. "As we were walking along," she recalled, "I noticed men lying all over the field. Some of them were motionless. Some were groaning, but nearly all of those that were lying down had their heads covered with blood and their clothing stained with blood." Steelworker Joseph Hickey may have been observing Marshall when he "saw a woman fall as she was being clubbed by the policeman. She was bleeding and looked like she was dying." Hickey, who was about 100 yards behind the front line of marchers when the assault began, ran to help the stricken woman, leaning down to pick her up. It was the last thing that he recalled from that chaotic scene, since he was then struck and knocked unconscious by police officers.[13]

Steelworker Louis Calvano wasn't a Republic employee, but he wanted to support the strikers on the Southeast Side that day. A union member himself, Calvano marched between the flag bearers and ended up on the front line, facing a phalanx of hostile police officers. Struck on the head, he turned and ran, but he was disoriented from the blow. Lee Tisdale, an African American resident of South Chicago, grabbed Calvano and pulled him along, helping him avoid the flailing billy clubs of the oncoming police assault. Traumatized by the initial blow he had received and dizzied by the tear gas, Calvano felt a bullet graze his cheek. He was only vaguely conscious of the fact that a moment before, Lee Tisdale had left his side. "I do not know if he was clubbed then or not."[14]

Reverend Fisk soon realized that the police had fired more than tear gas. He had forgotten about his camera, until he observed one young striker, suddenly motionless, drop like a stone. Fisk moved closer to the fallen man and snapped a picture of him, "lying with his face on the ground. I could tell he had been shot by the bloodstains on the back of his shirt." Looking up, Fisk saw police officers chasing marchers across the field and beating on them with their unusually heavy clubs. Two officers ran after one striker who pleaded with them to stop attacking him. "I'm going, I'm going," he cried. "I'm doing what you told me to do. I'm going as fast as I can." Running toward a nearby ditch, the young worker tried to jump it, but tripped and fell. The two officers approached him simultaneously. They "struck him down behind a little clump of bushes and then stood there for a couple of minutes slugging him." Swinging his camera into action, Fisk took several pictures of the officers hitting the collapsed striker "five or six times after he was down and apparently unconscious."[15]

Chicago Daily News reporter Ralph Beck, a witness to the verbal exchange before the attack on the young worker, also observed other policemen lunging at the strikers. "In two[s] and threes they pounded backs and cracked down on skulls. . . . Three policemen surrounded one falling man and while two held his hands and legs, the third kicked him in the groin. Two others stopped an old man with his hand held to his bald head trying to stop the blood, only to kick him and beat him across the back and tell him to hurry back to his mob." The elderly man could take only a few steps before he collapsed.[16]

Archibald Paterson, a motor-room operator at the Carnegie-Illinois plant and the financial secretary of SWOC Lodge 65, was one of the "ambulance" drivers. Pulling up alongside the marchers, Paterson was astonished to see police running down the retreating marchers, striking them from behind, and indiscriminately firing their weapons. "I saw one man who was coming on his hands and knees," Paterson reported. "He got up twice, and each time, as he got up, he was clubbed down again. The third time he lay still." Furious at what he had observed, Paterson grabbed a stick and ran toward the attacking police, only to realize the futility of his one-man assault. Thinking better of it, he headed back to his car. He would be of more help to his wounded comrades by ferrying them off the field than by facing down police revolvers.[17]

George Patterson had retreated quickly once the shooting broke out; he was lucky enough to avoid the tangle of fallen bodies directly in front of Captain Mooney. Stopping, Patterson observed volley after volley of tear gas fired at the marchers. He noticed that one of the canisters landing close to him did not explode, so he picked it up and hurled it back at the police lines, anxious to feel that he had somehow retaliated. "Then I saw the inert bodies of some people lying on the prairie grass. Others [sic] men were being clubbed unmercifully by policemen. A black policeman was down on one knee taking careful aim and shooting at us." With the acrid smell of gunpowder and the choking fumes of tear gas in the air, Patterson felt hostility welling up inside him. "I wanted to do something. Entice the police into the alleys. Get clubs, beat their brains out." Seething with anger, Patterson called on the other strikers to counterattack, but the bulk of them had quickly retreated from the police billy clubs. Dazed and disoriented, most of those who lingered on the field could not grasp the scope of what had just happened. "The cruelty I witnessed that day was forever etched in my mind," Patterson later declared. What he remembered most about those moments was the "wantonless clubbing, shooting of working men who were running away." Patterson thought he spotted policemen on top of the steel mill shooting at the marchers below. "I knew," he added, "that we had broken no laws."[18]

Uva Bohrte followed the events from the window of her Burley Avenue home, and she was transfixed by what was transpiring in the prairie. "It was horrible. I saw men and women falling everywhere, some of them shot and others clubbed. Many were bleeding terribly." Overcome by the tragic scene, Bohrte turned away, but her daughter Melsina took Uva's place at the window. "It was the most awful thing I ever saw," the *Chicago Herald and Examiner* quoted her as saying. "I saw scores of people lying on the ground and being lifted into ambulances, wounded and soaked by their own blood." Melsina believed that the experience would give her nightmares for years. "I never imagined," she added, "that human beings could fight and kill each other that way."[19]

The Injured

On the field, union supporters desperately worked to collect the dozens of wounded marchers who could not make it back to Sam's Place on their own. Overcome by his injuries and by the tear gas, Harry Harper found himself lifted into one of the impromptu ambulances, along with several other wounded strikers. They never got to Sam's Place. As soon as the driver had shoved the car into gear, several police officers jumped on the running boards and brought the vehicle to a halt. Pulling Harper and the other wounded marchers out of the car, the policemen threw them into a patrol van. "And then," Harper recalled, "that long journey started." With his left eye dislodged from its socket and his head bleeding profusely, Harper could not be sure if there were one or two other marchers who lay on the floor of the paddy wagon. One of the demonstrators sitting beside Harper was suffering from a gunshot wound to the thigh. The man pleaded with the officers to provide first aid; according to Harper, one of them contemptuously slammed the door, shouting "Shut up, you son of a bitch, you got what was coming to you." The policeman who was driving took a circuitous, seemingly directionless route to the hospital. A wounded marcher, still conscious and sitting beside Harper, called for the driver to stop at a doctor's office, since two of the marchers still showed signs of life. Harper was not sure whether the wounded striker was speaking about him.[20]

At the same time that some of the police herded Harper into one police van, others threw Lupe Marshall into another. The bluecoats made no effort to distinguish between the slightly and seriously wounded, hurling them all into the police vehicles alongside those merely suspected of being ringleaders. Cramming 15 and 16 marchers into poorly ventilated, overheated patrol wagons designed to hold only 8 people, the police further endangered those who were already in a precarious condition. "I had one foot on the step when a policeman put his hand on my back, on my buttocks, and shoved me in there," Lupe Marshall recalled. "If I had not put my hands across my face I would have struck the grating of the window in the front of the patrol wagon in there." Moving next to the door, Marshall observed the police roughly handling the severely wounded strikers. The police "started bringing them in by their feet and their hands, half dragging them and half picking them up." Turning to a 200-pound, "heavy-set" striker—probably Joseph Rothmund— the police "grabbed him, and shoved him in the wagon . . . and I noticed that he had two red stains, about the size of a penny, one on the upper side of his abdomen and one lower," Marshall reported. Sitting on the bench, she had to lift her legs to allow the injured workers to lie on the floor. None could sit up. The police "piled them one on top of the other. Some men had their heads underneath others. Some had their arms all twisted up, and their legs twisted up." By that time, 16 other marchers, all of them seriously wounded, had been stacked in the police van.[21]

After a seemingly interminable delay, the paddy wagons finally drove off. One patrolman rode on the back, to make sure none of the injured escaped. What appalled Marshall and the other demonstrators who had been corralled into police vans was how long it took to get to the hospital. "It was ages before we were able to get there, and every time the patrol wagon jolted, these men would go up about a foot or so, and fall on top of each other, and there was the most terrible screaming, groaning, and going on in that wagon!" Marshall tried to attend to the wounded strikers by shifting them into positions that would allow them to breathe freely. She discovered one injured marcher, Kenneth Reed, who was pinned underneath a heavy-set man and uncomfortably doubled over. Marshall managed to straighten him out and place his head on her lap. She quickly realized that he was in critical condition. His face was "getting cold and was black, turning black" as he motioned to a package of cigarettes in his shirt pocket. Retrieving the smashed-up pack, the social worker discovered that it was soaked in blood. "Never mind, kid," the wounded marcher feebly responded. "You are alright. You are a good kid. . . . Never mind. Carry on." Deteriorating quickly, he struggled to call out for his mother, but gasped helplessly instead. Until that moment, Marshall had been remarkably stoic, ignoring her own wounds and the swirling chaos around her. But something inside her finally snapped. "I hope you get the medal for this" she screamed at the patrolman riding on the back of the wagon. "Your children and your wife must be very proud of you." He was startled, but the Hull House social worker was probably just as surprised by his response. "I didn't do that," he insisted. "I wouldn't do that. I am just doing here what I can for you now. I am trying to help you as much as I can. That is all I have to do, is to see that you get medical care now." Contrite and apparently moved by the enormity of what had just taken place, the officer pleaded, "But I wouldn't do that."[22]

Still on the field, Earl Handley struggled to survive a gunshot wound to the leg that had lacerated his femoral artery and vein. One of the retreating marchers shouted to Archibald Paterson, who was frantically gathering up the wounded, that there was a badly injured striker stranded behind the advancing police assault. Just as his assistant, John Jablonski, jumped on the running boards of the makeshift ambulance, Paterson hit the gas and drove directly through the police line. When they arrived, they found two demonstrators carrying Handley in a blanket. "At a glance, I saw the man was in a bad way," Paterson stated. "His right leg was saturated with blood from the hip to the ankle. From the appearance of his face, he was losing blood rapidly." Paterson wasted no time in trying to fashion Handley's belt into a tourniquet in order to arrest the bleeding from the punctured artery in Earl's right thigh. Realizing how little time he had, Paterson and Jablonski moved Handley to the car and threw a blanket over him.

Paterson was just about to return to Sam's Place when a police officer pulled his weapon and ordered him to stop. Another policeman purportedly produced his service revolver and announced, "We will get one [of] them sons of bitches anyway." Paterson tried to convince the officers of how grave Handley's condition really was; without adequate attention, he could bleed to death in minutes. The cops then pulled Handley as well as two other wounded marchers out of Paterson's vehicle; one of these was Harry Harper. Shocked at this callous indifference, Paterson appealed to the policemen to place Handley in the blanket. "They swore at us again," Paterson bitterly recalled. "Four constables got hold of the man," one grasping the tourniquet that Paterson had struggled to get into place. It slipped uselessly down to

Handley's knee, and Paterson saw the blood flowing freely from the top rather than from the leg of his pants. They dragged Handley to the patrol wagon, where he would join Harry Harper in that "long journey."[23]

While marchers fled or crumpled under the crack of swinging billy clubs, the Paramount newsreel photographer kept a steady hand on his camera. After viewing this astonishing footage, journalist Paul Y. Anderson of the *St. Louis Post-Dispatch*, who had made his name uncovering police complicity in the murder of African Americans during the East St. Louis race massacre of 1917, wrote the following description of the scene. First, "a terrific roar of pistol shots" is heard suddenly, inexplicably, followed by people in the front rank of marchers collapsing like "grass before a scythe." The fusillade of gunfire generates a "massive, sustained roar" for about two to three seconds. The camera then captures the police charging at the strikers. While tear gas canisters sail through the air and strikers look wildly for escape routes, the police launch a nightstick assault in an "appallingly business-like" manner. Clutches of police officers "close in on these isolated individuals, and go to work on them with their clubs." In several instances, multiple police officers are seen beating one man. "One strikes him horizontally across the face, using his club as he would a baseball bat. Another crashes it down on top of his head and still another is whipping him across the back." The scene then changes, as the camera zeroes in on a striker—probably Otis Jones— "shot through the back" and "paralyzed from the waist. Two policemen try to make him stand up, to get him into the patrol wagon," the reporter continued, "but when they let go of him, his legs crumble, and he falls with his face in the dirt, almost under the rear step of the wagon." The wounded man pathetically moves his head and arms, but his legs are completely unresponsive. "He raises his head like a turtle and claws the ground." The stream of conversation is unintelligible, and the film registers only two significant sounds. The first is the thunderous volley of revolver fire. The second is equally prominent, yet surprising, since it emerges so distinctly out of the welter of voices following the attack. It's almost as if everyone else briefly went silent long enough to hear it. The speaker is a police officer, but we do not know which one. What we do know, as Anderson notes, is what he said: "God Almighty!"[24] Evidently not all of the police officers were indifferent to the tragedy that had just played out on a desolate field in Southeast Chicago.

Congregational minister Chester Fisk was another observer who found himself hauled in by the police dragnet that followed the attack. Herded into a patrol wagon, Fisk discovered 11 other marchers there, 4 of them seriously wounded. Two demonstrators "had their heads laid open in several places so you could hardly see their hair for the blood on their heads." The two other injured strikers had been "beaten so nearly unconscious that they were sitting on the seats of the patrol wagon in a daze, rocking back and forth," barely able to sit upright. Each wore a shirt saturated in blood that flowed freely from untreated wounds.[25]

Steelworker Louis Selenik had been clubbed repeatedly by the time he was thrown into the paddy wagon carrying the Reverend Fisk. Selenik discovered that the wounded in the patrol van included Jayson Johnson, an African American; Max Guzman, one of the flag bearers; Nick Krugar, a steelworker of Eastern European descent; and Ada Leder, a member of the Women's Auxiliary who was five months pregnant at the time. They represented a cross-section of the Popular Front movement that propelled the steelworkers' campaign. "I was bleeding," Selenik recalled. "Nick Krugar was bleeding. The girl was bleeding." As Selenik recalled, Krugar had been shot in the head and clubbed in the back. Selenik remembered looking out of the window of the paddy wagon and seeing "one fellow that was laying down with a great big hole in his abdomen that was terrible.'" Selenik couldn't be certain, but he believed at the time that one of the demonstrators died in the patrol wagon as it made its way to the South Chicago jail. According to Selenik, police held the wounded and dying there for over an hour before transporting them to the hospital.[26]

The situation that Dr. Lawrence Jacques confronted at Sam's Place was no less horrific. Sympathetic to labor, the young physician was a typical supporter of the labor-led progressive movement. He was a University of Chicago graduate with a cosmopolitan worldview, formed by study in Vienna, Paris, Edinburgh, Germany, and Switzerland. Jacques had been at the temporary strike headquarters on Thursday and Friday, May 27 and 28. He had treated some of the wounded who limped back to Sam's Place after the fracas on Friday evening. On Sunday, he thought he had come prepared. Equipped with iodine, cotton, and gauze, Jacques probably expected nothing more than the scalp lacerations and soft-tissue contusions that most of the strikers had sustained on the earlier march to the Republic Steel plant. The young physician watched the marchers assemble on Memorial Day and followed them toward the prairie, trailing the end of the line by about 100 yards. He observed the marchers stroll toward the police lines and stop. Just as Jacques turned around to go back inside Sam's Place, he heard a "report which sounded to me almost like machine-gun fire," followed by a quick series of explosions. First blue smoke appeared, then white. What Jacques observed next was surreal. The crowd pivoted, almost like the "slats of a Venetian blind." The marchers seemed to be rushing off the field, returning in the direction from which they came, apparently heading toward Sam's Place.[27]

Jacques quickly surmised what was in store and rushed back to his makeshift field hospital. "Within 4 or 5 minutes, there were approximately 30 or 40 bleeding, groaning, screaming, dying, and I thought one dead person." Jacques had the able assistance of several union supporters, including Women's Auxiliary leader Virginia Mrkonich. Since the beginning of the strike, Mrkonich had worked to recruit volunteers and gain the support of local businesses on Chicago's East Side. When one of the merchants failed to support the strike, she informed him that he could expect a boycott. Busy at Eagles Hall—the official strike headquarters—when the march began, Mrkonich rushed down to Sam's Place to serve as a medic to the victims now flooding into the former tavern. She and the other ad hoc nurses retrieved white shirts from the wounded strikers and converted them into bandages. Knocking on doors in the neighborhood, she solicited pillowcases and sheets to use as dressings. While Mrkonich attended to the wounded in Sam's Place, Women's Auxiliary member and steel activist Lucille Koch began transporting the critically wounded to the South Chicago Hospital. Koch had been a key figure in the Friday demonstration at the Republic Steel plant. A photo of her lending assistance to a wounded striker while carrying the American flag became an iconic image in the early days of the struggle. Now, ignoring the railroad-crossing gates at Wisconsin Steel, she sped her vehicle toward the hospital.[28]

When Marilee Kone and the other students returned to Sam's Place, more than 50 wounded had arrived from the field. "During this whole time," Kone observed, "the crowd was amazingly quiet." Few spoke, except to order those in the tavern to "clear the way" as another wounded striker arrived. One of the injured was James C. Row, a Youngstown Sheet and Tube worker. Knocked unconscious in the attack, he was struck again the moment he stood up; taking flight, he dodged yet another officer who had zeroed in on him. "He just missed hitting me," Row recalled. "If he had struck me, I think he would have knocked my brains out." Bleeding profusely and hopelessly disoriented, Row would have collapsed but for the efforts of two other marchers who grabbed him. "You are alright now," they told him as they whisked him off to Sam's Place. Kone and four of the other theology students then courageously returned to the scene of the attack. It was easy to find precisely where the onslaught had occurred. The "ground was torn up," Kone noted, "and spotted with blood." When the students tried to interview the police about the horrific clash, they were told to return to union headquarters and "ask the 'communists' who started the trouble."[29]

Making a quick assessment of the mounting casualties, Jacques discovered between 15 and 20 gunshot wounds and dozens of lacerations. Well aware of his limited resources and personnel, Jacques and two other doctors decided to transport the wounded to local hospitals. James C. Row, who was huddled into a car with two or three other wounded SWOC members, noticed his friend, Joe Hickey. It was Row who had driven Hickey to the meeting earlier that day. Hickey's "head and shirt was covered with blood and he was laying to one side." Shortly thereafter, a call came in for additional medical assistance at South Chicago Hospital. Jacques obliged. Arriving there, he found the hallways jammed with the injured, interns, nurses, and police officers. In the emergency room, two men seemed to be gravely wounded: Hilding Anderson and Anthony Tagliori.[30] The latter was the severely injured marcher that Louis Selenik spotted on the ground outside the patrol wagon.

The protracted trip to the hospital was almost as traumatizing as the vicious assault on the field. Harry Harper recalled appealing to the police for first aid to relieve his agony. "Shut up, you son of a bitch," one of the officers responded. "You got what was coming to you." As Harper faded from blood loss, another marcher, shot in the leg, appealed in vain for the police to stop at the nearest hospital in order to help those who were gravely wounded but still breathing. Harper remembered thinking that the injured demonstrator was referring to him. After an inexplicably long trip, the van finally arrived at Bridewell Hospital. By that time, Harper was "terribly weak." He noticed that he was bleeding on the people lying at his feet. The police commanded Harper to get up, but he was in no condition to comply. Undeterred, they dragged him out of the paddy wagon feet first, his head hitting the steps on the way down. According to Harper, the officers chose to take him in for medical attention but decided to deliver the others in the police vehicle to the morgue, even though no physician had pronounced them dead. Ben Mitckess, a steelworker at Youngstown Sheet and Tube, could not believe his eyes as the police extracted Harper and the others from the patrol van. The police sat Harper down without any medical attention until some hospital personnel removed his clothes, leaving him "cold, freezing, and shivering."[31]

An equally surreal scene played out at Burnside Hospital, where Lupe Marshall and 16 others arrived that evening. The police unloaded the wounded as carelessly as they had thrown them in, hauling them out by the feet and hands and leaving them on the hospital's concrete floor. Marshall did everything she could to help the more seriously wounded. As the few available nurses rushed to attend to the arrivals, Marshall collected tablecloths, napkins, and a pitcher of water from a nearby dining room. Using these crude items, she improvised wet packs to apply to the wounded. Just as she was offering assistance, a detective burst into the emergency room and made "a terrible noise." He yelled at the officers standing in the doorway, demanding to know who ordered the shooting. Some of the police began to respond, but the officer assigned to watch Lupe Marshall intervened. "Shut up your mug!" he growled. "They are not all dead yet." Marshall continued to try to help the victims.[32]

She also finally received treatment for her own wounds. Bandaged and suffering from the combined effects of heat exhaustion, shock, and blood loss, Marshall was sent for x-rays. The police followed. Arriving at the x-ray floor on the elevator, she was surprised to discover an officer waiting for her. When Marshall entered the washroom, the policeman rushed in and grabbed her, dragging her into the hallway and down the stairs. Pulling her out onto the hospital's driveway, he demanded that she give him her name. The social worker from Hull House repeatedly insisted that her lawyer was on the way. According to Marshall, that was enough to enrage the constable. Although he threatened to strike her with his billy club, a frightened Marshall kept her wits about her. "You can't do that," she shrieked, "because you officers have to be more within the law than ordinary citizens." That was enough to deflect the attack, but not enough to prevent Marshall from being hauled off to the Burnside police station, "feeling very ill." There, police repeatedly questioned her. They rifled through her purse and pulled out two pieces of literature, one announcing the protest meeting at Sam's Place, and the other, a U.S. Post Office auction. "Communist stuff," the police matron reputedly announced. It was only on Tuesday night, June 1, that Marshall was finally charged with conspiracy to commit an illegal act. She had yet to see her lawyer.[33]

Harry Harper, who was among the most seriously wounded of those who had survived the Memorial Day massacre, captured the sentiments of the majority who marched on May 30. "I was not armed," he claimed. He even came dressed in his best clothes for a festive protest. He did not see a mob of demonstrators at Sam's Place armed with guns and clubs. Had that have been the case, he "would not have gone over. I have a wife and child and there was nothing for me to gain by going there if there had been trouble." Like many of those who testified about what they experienced that day, Harper recalled seeing women and children in the parade, as well as men with their wives or girlfriends. Their presence offered ample testimony about the nonviolent character of the march. "There was no intention of trouble." Otherwise, why would a person bring his wife, son, daughter, or parent along?[34] Quite simply, the police had wielded lethal force against a mass demonstration that had little in the way of self-defense and even less in the way of aggressive intent.

Less sympathetic observers told a different story. Reporting for the virulently antilabor *Chicago Tribune*, Edwin J. Kennedy later claimed to have seen strike supporters carrying "clubs, iron bars, bricks, and slingshots" prior to the march. According to Kennedy, he saw several of these men practicing their slingshot delivery and lining up in "company formation" while Nick Fontecchio addressed the crowd at Sam's Place. Positioned behind police lines during the march, Kennedy claimed to have heard Captains Mooney and Kilroy pleading with the marchers to leave the field peacefully. Kennedy maintained that the officers calmly but firmly notified the protestors that they were "assembling in violation of law." While some demonstrators chanted "CIO, let's go," others allegedly fired a barrage of bricks and projectiles at police ranks. Instead of the "backward motion" that Reverend Fisk observed, Kennedy stated that he saw the rear ranks of marchers pushing against the forward line. Then he heard two gunshots. According to Kennedy, they originated from the middle ranks of protestors. "Some of the marchers were seen raising their clubs," the unusually observant Kennedy noted. In quick succession, policemen lobbed tear gas canisters and fired their revolvers at the crowd. "I then heard about two hundred shots fired," Kennedy reported. In his opinion, approximately 20 of the officers fired into the air.[35] That statement raises a question: did some of the police misinterpret a initial warning shot, or several warning shots, fired into the air as a cue for directing a volley at the crowd?

Kennedy and a number of police officers would testify that the demonstrators hurled more than insults before the police assault began. Sergeant Lawrence Lyons claimed that the marchers pelted the police ranks with rocks and pieces of concrete. Then the demonstrators attacked, using tree limbs, jack handles, and railroad ties against the police. "I was struck with a rock in the stomach," Lyons claimed, while the billy club he was holding split when it was hit by a flying projectile. "The first thought that came to my mind," he recalled, "when the rocks came through the air was my wife and four children—that is all."[36] This would become the standard police version of the encounter: an intense "shower" of missiles launched by the protestors precipitated a response by the police, who had little choice but to defend themselves and the steel mill. Their take on the May 30 events would carry at least some credence, since one of the photos snapped that day by journalists clearly showed some police officers cowering from what they would later claim to be a deluge of projectiles. The problem with this interpretation is that the same photograph unquestionably showed not only the "backward movement" described by Frank McCulloch and similarly mentioned by Reverend Fisk, but a ragged line of marchers in full retreat.

Officer William H. Cannon provided an even more vivid description. What he observed was a "mob" steadily and menacingly moving toward police lines. In Cannon's account, the protestors had armed themselves with meat hooks, brickbats, stones, iron pipes, jack handles, baseball bats, broken bottles, and other menacing objects. "They were cursing and yelling at the police and yelling and telling them what they were going to do to them and the men in the Republic Steel Company's plant," Cannon remembered. After hurling the projectiles, the marchers attacked, "swinging their clubs and weapons." After the marchers had allegedly fired shots "from the center of the mob," they allegedly hurled another wave of stones and brickbats at the police defenses. Captain James Mooney further embellished this lurid reconstruction of events. Mooney asserted that after reading his announcement aloud, a striker brandishing a club with a meat hook attached threatened to "put that through your skull," while a marcher who "looked like a Mexican" spat on Lieutenant Ryan. "Then we got a barrage of bricks, all kinds of missiles, and I told Lieutenant Moran to throw the gas bombs, so he threw them." In this version of the Memorial Day events, only then did the police, driven by self-defense, retaliate against the marchers.[37]

George Higgins, the patrolman who confronted Lupe Marshall, provided an account that closely followed the official police version of the story. According to Higgins, "the mob" was determined to enter the Republic Steel plant. He claimed to have observed bricks and ball bearings hurtling toward

"Police in pursuit of retreating marchers as tear gas wafts up from the field," Exhibit 1439 in *Violations of Free Speech and Rights of Labor: Hearings Before a Subcommittee of the Committee on Education and Labor, United States Senate*. Taken from the right of the police line, this photo captures a couple of police officers crouching, perhaps to avoid what some observers claimed were objects thrown from the back ranks of the marchers. Note that at least a few strikers are retaliating against the police, a response that would not have been at all surprising to observers during that era. The overwhelming majority of marchers are in retreat across the field, moving toward Sam's Place. The Chicago History Museum, used by permission.

the police lines. Higgins, standing to the right of Captain Mooney's position, claimed to have heard shots fired. He then observed "one lousy pup with a sawed-off shotgun," apparently firing scattershot and BBs. When Officer George Barber went down, wounded by a slingshot projectile to the jaw, Higgins asserted that he saw "this lousy pup coming with the gun at me and I had lost my club in this scrap . . . and I waited for him to get set, and then . . . I smacked him." Higgins had been exceedingly busy during the altercation. He had allegedly spotted Officer Walter Oakes pinned to the ground by a striker whom Higgins claimed had his knee pressed into the officer's neck. Higgins then saw a "second mobster with a short club about two foot long, shaped like a wedge, and he's raising the club. . . . And I wait for my chance and measure him off and sock, I smacked him." Having been freed, Oakes was able to bound to his feet and get back into the action. Oakes "shot this Rothmund, who we identified at the morgue, and Oakes shot him again and perforated him in the stomach."[38]

Higgins admitted that he shoved Lupe Marshall. But the officer contended that when Marshall fell, she dropped a makeshift weapon designed to disable the police officers. Despite Higgins's claim that "I picked up a bag of pepper that she [Marshall] had and turned it in," the bag was never submitted as evidence and never appeared in any subsequent testimony. Neither did the handguns which Higgins and other officers claimed the marchers had carried. Not a single firearm was discovered on the field or confiscated from a marcher taken into police custody. The police did not charge any demonstrators with carrying concealed firearms. As for the cars bearing red crosses, which served as ambulances to carry the injured demonstrators from Sam's Place to area hospitals, Higgins acknowledged that he turned back two of them. "Get out of here, you goddamned rat," he admittedly shouted at one of the drivers. Yet Higgins also declared that the conflict on Memorial Day eclipsed anything he had experienced. "I been in that race riot [of 1919], I been in that stockyards strike with all them foreign savages out there, but this beats them all." Higgins's attitude toward immigrant workers was clear enough. What disturbed him most, however, was that these "foreign savages" had joined working- and middle-class women—as well as African American, Anglo, and Mexican steelworkers—in a direct challenge to his authority. Clearly, these people no longer knew their "place."[39]

The Day's Final Tally

By evening on Memorial Day, a total of four of the marchers had died. Sam Popovich, 50, had been beaten by police and killed by a gunshot wound to the head. Kenneth Reed, 23, was shot three times, fatally struck by a bullet to the back. Earl Handley, whom Archibald Paterson and John Jablonski had struggled desperately to help, passed away that afternoon. "The effective application of a tourniquet," Dr. Jacques reported, "could certainly have saved his life . . . the effective and prompt application of a tourniquet," which was precisely what Archibald Paterson had tried to convince the police to help him do. Alfred Causey, a 43-year-old steelworker originally from Alabama, also died that day, shot four times and beaten so viciously that Jacques identified "extensive lacerations on the scalp and back of the head."[40] The Paramount newsreel photographer captured Causey in his final moments, dragged along on the side of the dirt road like little more than the detritus of battle. That same footage also portrayed at least one police officer trying to alleviate Causey's suffering by placing a placard under his head for support. If most of the police had behaved like uniformed goons that afternoon, at least two or three had shown a spark of humanity.

The affiliations of the casualties testified to the extent of the working-class coalition of marchers that day. Those killed on Sunday, May 30, had all been steelworkers from Indiana Harbor; each had come to support the strikers at Chicago's Republic Steel plant. The next day, Joseph Rothmund slipped away. He had sustained a bullet wound and had undergone surgery at Bridewell Prison Hospital, but to no avail. Anthony Tagliori died as well on June 1, pierced in the back by a bullet that penetrated his bowel and bladder. He ultimately succumbed to peritonitis. On June 3, Hilding Anderson, a worker at US Steel's Carnegie-Illinois plant, expired from a gunshot wound to the side that was inflicted on Memorial Day. On June 8, 33-year-old Otis Jones of Hegewisch, a community on Chicago's South Side, died from the pneumonia he contracted while trying to recuperate from a bullet wound to the back. The slug had paralyzed him from the chest down. Leo Francisco, age 17, lingered for two weeks, finally dying on June 15 from blood poisoning that set in from a gunshot wound in his thigh. Dr. Jacques interviewed Francisco before he died and later examined the forensic evidence, concluding that the bullet had initially entered Francisco's back. Lee Tisdale, the African American worker and activist from Chicago's South Side who had helped Louis Calvano escape police billy clubs, died on June 19 in St. Luke's Hospital. He was the tenth fatality of the Memorial Day assault.[41]

In addition to the marchers who were killed, Dr. Jacques counted 30 additional victims of gunshot wounds resulting from the bloody Memorial Day confrontation. By his tally, 28 others suffered injuries severe enough to require hospitalization, 8 of which seemed to threaten permanent disability. Another 25 to 30 sustaining major injuries never saw the inside of a hospital. In total, the police assault wounded between 100 and 110 marchers, 10 of them fatally. By comparison, the police suffered only superficial wounds. This stood in stark contrast to the kind of injuries that Lupe Marshall saw at Burnside Hospital, where one marcher arrived with his "head opened in five places" and a dent on his forehead so deep that the physician could place his finger in it.[42]

A firestorm of controversy was just about to erupt over who was responsible for the killings. What should be clear to historians is that the Memorial Day incident can only be understood in the light of the events of the previous week. It needs to be examined in the context of the working-class militancy that had been on the rise since the Unemployed Councils' first protests in the early 1930s. The march on Chicago's Republic Steel plant was part of the larger "structure-shifting matrix event" that developed during the Great Depression and reached a crescendo during the wave of sit-down strikes in 1937. The Memorial Day demonstration was not an anomaly, so we should not be surprised to discover that some workers came prepared for a confrontation, and that some of them did indeed fight back, however ineffectively, once the police riot began. Nor was the bluecoats' response inexplicable, considering how earlier protests, virulent antiradicalism, and the legacy of racial and ethnic intolerance shaped their worldview.

By no means do the day's events imply that the marchers had precipitated the police assault, or that the credibility of the drive for unionization was compromised by the willingness of the workers and their supporters to engage in a mass demonstration. What they suggest is that the Memorial Day march to the Republic Steel plant on Chicago's Southeast Side was part of the fabric of the movement for social democracy that broke out during the 1930s. This was not a story of innocent (or naïve) bystanders and demonic police officers. Instead, it was the most conspicuous evidence that class conflict had come to industrial America. The demonstration itself was intended to be nonviolent. Marchers did not attack when they encountered the police cordon, and they had not planned to assault the nonstriking workers on the inside of the plant. Even so, we should not delude ourselves: the demonstrators *had* engaged in a direct-action protest, the kind that has often elicited a violent response from those in positions of authority. More than this, there is sufficient testimony from both sides to make the case that some steelworkers, however small in number, did at least discuss the possibility of marching into the plant that day. Les Thornton, who worked at Youngstown Sheet and Tube in Indiana Harbor and supported the strike, later acknowledged that the steelworkers had prepared for a confrontation that day. "I was quite aware that we were preparing for violence because we had not only people who were going down there for the purpose of attending the meeting, but we had organized first aid squads and arm bands and all of this business, which only spelled one thing to me." Thornton believed that, had the marchers not been stopped by the police, the demonstration would "probably have resulted in our going in to the plant and tossing out the scabs. *I really didn't see anything wrong with it*" (emphasis added).[43] On Memorial Day 1937, the mass demonstration had been turned back, but at an extremely high cost. The question now was whether this would galvanize the liberal/labor alliance in Chicago, or completely derail the strike.

4 Red Scare and Popular Resistance

BEFORE THE FUNERALS for the first victims took place, the media and the police had launched a Red Scare that would revive memories of the anarchist Haymarket episode of 1886. In the events leading up to that notorious nineteenth-century incident, Chicago police shot and killed four striking workers at the McCormick Reaper factory, and local anarchists planned a rally at Haymarket Square to protest the brutality. When someone threw a bomb, killing at least one officer, the police responded by firing recklessly into the crowd, possibly inflicting mortal wounds on seven more police officers that night. The conflagration wounded an additional 67 officers. Indiscriminate police shooting killed 4 workers and injured another 50. Convinced that the anarchists had spearheaded a much larger revolutionary conspiracy, the police and local authorities launched a manhunt that rounded up any and all labor activists they could get their hands on.

This dragnet eventually netted eight suspects; only two of these had been at the square when the bomb detonated. The state initially won convictions against all the defendants, with the preponderance of the cases resting on evidence that the accused had espoused anarchism. Three of the eight eventually had their sentences commuted, and a fourth managed to have explosives smuggled into the jail, which he then used to blow himself to bits. The remaining four died on the scaffold, but not before August Spies, his hands and feet bound, cried out to the witnesses: "The time will come when our silence will be more powerful than the voices you strangle today."[1] This was the same Spies who, at the trial, declared that the labor movement had become a "subterranean fire" which would continue to burn beneath the feet of the powerful, despite their best efforts to stamp it out.[2] Was Spies right? Were these silenced voices from 1886 now speaking through the labor upheaval of the 1930s? Was the Little Steel Strike the "subterranean fire" to which Spies referred? And would the combined forces of police and government coercion, aided by media complicity, be enough to stamp it out completely?

In 1937, there is little doubt about how the *Chicago Daily Tribune* saw the Memorial Day incident. It quickly became a zealous voice for law and order, Chicago style. More than this, it supported the efforts of the steel industry, American businesses, southern conservatives, and anti-Roosevelt Republicans to derail the New Deal. Those who had joined the united front in Chicago viewed it differently. For the intellectuals, reformers, and many of the frontline steelworkers who held the picket lines, the struggle in Chicago had become one phase of the larger contest at the center of this decisive decade. The steel movement had become fused with the wider movement for social democracy in modern America. "Civil liberties," the *Daily Worker* reasoned, "if they are to have any meaning [in] Chicago, must be fought for by a united coalition of progressives and labor groups." "At present," the editors added, "civil liberties do not exist in Chicago. The Constitution is a dead letter."[3]

By June 2, the city's daily newspapers had fabricated a vast communist conspiracy. The *Chicago Daily Tribune* reported that Margaret Rothmund, wife of Joseph Rothmund, one of the marchers killed in the field on May 30, admitted that her husband had belonged to the Communist Party; even more incriminating, he kept "communist literature" in the house. Rothmund had been photographed distributing the *Daily Worker* and arrested for taking part in action by an Unemployed Council to prevent an eviction in 1934. The paper also vindictively reported that Rothmund "was not a steelworker, lived nowhere near the plant, and was accepting the government's pay as a Works Progress Administration employee." The insinuation was that he was an "outside agitator." The idea behind the articles underlining Rothmund's involvement in the Works Progress Administration (WPA), a New Deal

A group of women attend the funeral of workers shot and killed during the May 30, 1937, march to Republic Steel's Burley Avenue plant. Women had played a decisive role in the organizing drive and during the first week of the steel strike. In the aftermath of the massacre, they would lead protests against police brutality while trying to maintain the picket lines. Ben and Beatrice Goldstein Foundation Collection, Library of Congress, courtesy of David Lubell.

agency that provided hundreds of thousands of jobs for the unemployed at the height of the Depression, was to implicate the Roosevelt administration in the alleged labor violence of Memorial Day. The *Chicago Daily Tribune* also informed readers that the CIO lawyer who cross-examined Margaret Rothmund had defended communist leader Earl Browder when he was arrested in Terre Haute, Indiana. Guilt by association was piled on top of unfounded accusations, and everything was bound together by the antilabor prejudices the mainstream media had been encouraging for years. Soon the newspaper would simply refer to Rothmund and his comrades as the "Reds" responsible for the "riot."[4]

Police captain James Mooney fuelled the conspiracy theory by announcing that he had evidence of communist responsibility for the May 30 march. Echoing the newspaper's assertions, he concluded that, since the Illinois state committee of the Communist Party had produced pamphlets so quickly after the incident, the committee members must have had prior knowledge of the supposedly planned attack at the Republic Steel plant. It apparently never occurred to the captain that the efficient production of this kind of boilerplate propaganda was the party's stock in trade.

Yet Mooney and Coroner Frank J. Walsh soon had additional intelligence. On Tuesday, June 2, Make Mills, an investigator with the Industrial Detail of the police department's Red Squad, reported that 13 of the marchers from Sunday's demonstration belonged to the Communist Party. The Red Squad had been busy throughout the 1930s, infiltrating and spying on left-wing organizations, particularly those attempting to organize workers. Now it reported that Ada Leder, five months pregnant and the wife of a CIO member, was a party member. The newspapers labeled Paul Tucker, John Telick, and Joe Starcevuck as communists because they had joined left-wing unions, participated in public demonstrations, and belonged to the Unemployed Councils. Lupe Marshall, a social worker and key witness to the brutality on Memorial Day, was also accused of being a party member. Authorities believed that her participation in a "parade demonstration" in 1935 was evidence enough of her communist affinities. The same reasoning was applied to Leder, who appeared in a group photo of a suspicious left-wing organization. Under questioning by police at the Eighth District Station only hours after the march, Leder herself supported the accusation that some of the demonstrators wanted more than a token march. According to Leder, "they wanted to walk through the plant and get the boys to come out of the plant."[5]

That may very well have been the case for a minority of the protestors. Considering the diverse composition of the group and the prior's week's frustrating effort to establish a picket line, it would not be surprising if at least a few had entertained the idea. But with a considerable number of women and children in tow, the notion of a mass invasion of the Republic Steel plant was far fetched at best. By drawing a straight line from communism to labor activism and political violence, the authorities sought to discredit the entire labor movement in the steel industry. Their antiunion strategy was to emphasize the political and ethnic *identities* of the victims, while downplaying or simply ignoring the *behavior* of the aggressors. Newspapers and the political establishment did more than implicate the Communist Party in labor violence, however. They wanted to erase some of its prominent stances from the public's consciousness: its advocacy for racial equality, its defense of the rights of homeowners and tenants, and its commitment to the needs of the unemployed and the destitute.

Some of those on Make Mills's list certainly did belong to the Communist

Party. WPA worker Joseph Rothmund, shot and killed during the Memorial Day march, was a party member. So, too, was Hank Johnson, an African American who was active in the left-leaning National Negro Congress and was a SWOC organizer. There is good reason to believe that Ada Leder was a party member, and Lupe Marshall certainly had no qualms about working closely with party activists. In fact, the Communist Party was known to use Hull House for some of its meetings, and Hull House was Marshall's base for social work in the Latino community. Yet what did all of this mean? Despite their fiery rhetoric about class conflict, American communists never did champion political violence. In the era of the Popular Front, they defined themselves by their support for the CIO, the social democratic agenda of the Roosevelt administration, and the principle of racial equality. The Communist Party opposed fascism abroad and at home and, at least during those years, downplayed the idea of revolution. Instead, the party emphasized the importance of building antifascist alliances with liberals, socialists, and other fellow travelers. What concerned Chicago's media and the city's political elites the most was not Stalinism. It was the party's effectiveness in organizing steelworkers in 1937, a year of unprecedented labor rebellion.

For the police, the Memorial Day Massacre was not simply a labor dispute that turned ugly, but a threat to the existing order. According to Captain Thomas Kilroy, it was the force's "duty to protect life and property and I am sure that those same fellows who are criticizing us today would be the first to call on the police for help if a mob was surrounding their homes and they thought that their lives and the lives of their families would be endangered." The protection of property and life became the mantra for the police department. Cook County State's Attorney Thomas Courtney reiterated that position. "Property rights will be protected," he insisted. "We will track down those responsible for this riot and punish them to the full extent of the law." According to Courtney, property rights took precedence over national labor law and the constitutional rights of demonstrators. As Courtney's comments inadvertently suggested, the Little Steel Strike had developed into something more than a fight over union recognition. It was now a contest over whether the private-property prerogatives of industry had primacy over the human rights of American workers.[6]

According to Chicago's leading newspaper, the strike had pitted liberty-loving Americans against communist conspirators. By June 7, the *Chicago Daily Tribune* had linked the demonstrators with Moscow and international communism. It asserted that the "C.I.O. is working hand in glove with the Communist party in the current warfare on capitalism."[7] The paper's editors now claimed that communists had been in the front ranks of the "South Chicago battle." According to the *Daily Tribune*, Moscow had a friend in Roosevelt. It was the president who had permitted the sit-down strikes to flourish. It was the president who had "forced" General Motors to negotiate. It was the president who had allowed John L. Lewis of the CIO to "dominate" the Department of Labor, and Lewis was now demanding a political return on the CIO's campaign contributions to the Democratic Party. At this point in time, the mainstream media saw the conflict in Chicago's Southeast Side as a flashpoint in a much larger struggle.

A Persistent Movement after the Tragedy

Despite the loss of life and the savage media backlash, the drive for industrial unionism continued to build following the massacre. It was apparent in a SWOC-sponsored mass funeral for three of the marchers killed on Memorial Day. It was evident as well in a June 18 protest meeting at Chicago Stadium, which drew between 16,000 and 20,000 people. It began with an automobile parade through East Chicago, Indiana, and ended with an electrifying speech by Thomas Kennedy, the lieutenant governor of Pennsylvania and a former mine worker. At a meeting attended by CIO-affiliated rubber, packinghouse, leather, and textile workers, Kennedy proclaimed that "it is vital that labor organize and redistribute the wealth of this nation."[8] America was "thinking differently these days," Kennedy announced. "Those steel barons don't realize it. America is thinking in terms of human values."

Additional protests demonstrated the willingness of Chicago's workers to challenge police brutality and defend the Wagner Act: 16 University of Chicago students joined 8 steelworkers in picketing duty at Burley Avenue and 118th Street, and 15 members of the Women's Auxiliary demonstrated outside city hall on LaSalle Street (in the Loop, the center of the city's business district). Led by auxiliary organizer and Communist Party member Minneola Ingersoll, as well as auxiliary member Regina Mrkonich, the women established a picket line featuring signs that read "Who Gave the Order to Shoot to Kill?" and "Mayor Kelly, Wipe the Blood Off Your Hands."[9] The day after the massacre, 5,000 workers turned out in Indiana Harbor to hear SWOC regional director Van Bittner and to rally support for "democracy in steel."[10]

The most evocative demonstration occurred on June 8, when the Chicago Citizens' Rights Committee held a protest at the Chicago Civic Opera House. The committee read like a "who's who" of Chicago's Popular Front. It included Robert Morss Lovett, a left-wing English literature professor at the University of Chicago who, as a General Motors stockholder, had written eloquently of the primacy of the rights of labor over the rights of property during the sit-down strike at that company's Flint, Michigan, plant. Paul Douglas, a liberal economist at the University of Chicago and erstwhile advisor to the Roosevelt administration, was also present. Attorney Alfred Kamin of the National Lawyers Guild, poet Carl Sandburg, labor activist A. Philip Randolph, writer Meyer Levin, Professor Earl Dean Howard of Northwestern University, and Professor Malcolm P. Sharp of the University of Chicago all joined in this demonstration of solidarity with Chicago's steelworkers. The meeting, which featured testimony by injured workers and shocking photographs of the Memorial Day Massacre, condemned police brutality and city hall complicity. It also demonstrated the youthful vitality of this movement to change the social and economic landscape of modern America. As Albert W. Palmer, president of the Chicago Theological Seminary and member of the Citizens' Rights Committee, observed, "the most significant fact about them was that they were all young—hardly a gray hair present in the audience. It was a great representative meeting of the youth, primarily the working class youth, of Chicago."[11]

The atmosphere in the opera house was electric as 4,600 supporters crowded in to hear the wounded marchers recount their Memorial Day experiences. The torn flags that John Lotito and Max Guzman had carried were displayed, providing vivid representations of the devastation inflicted on *their* vision of the American creed. "I am a striker," announced steelworker Emil Riccio as he stood before the pulsating crowd. "I try help my brothers. We start parade nice and peaceful. I was back in line. I saw men stopped by police then I heard shots. The shots came more fast than a machine gun. Then I got hit in shoulder and fall down."[12] Middle-class supporters also took to the stage, articulating the larger significance of the events and placing the police riot within that context. Robert Morss Lovett condemned Captain Mooney as a murderer; theology student and eyewitness Clayton Gill reported that the "police fired without provocation and continued to shoot while the crowd fled"; Meyer Levin, a former student of Lovett's, read a list of the dead strikers, and Dr. Lawrence Jacques offered medical commentary on

"Republic Steel protest at City Hall." A group of women—probably relatives or associates of steelworkers—supporting the strikers and those killed in the Memorial Day march. Demonstrating in downtown Chicago, they demanded justice from the mayor's office and the police department. Women proved to be indispensable activists in the liberal-labor alliance that supported the working-class movement of the 1930s. The Chicago History Museum, used by permission.

their demise. Van Bittner assailed Republic Steel for its unholy alliance with city police.

It was left to Palmer to articulate the connection between the Popular Front and the events at Republic Steel's Burley Avenue plant. On behalf of the churches and the community, Palmer apologized to the victims of May 30. He did so because these entities had "failed . . . in adjusting the deeper background tensions between labor and management in our industrial life. It is because we are so selfish," he lamented, "stupid and self-willed that we have not yet learned how to give labor a fair, recognized and democratic way in which to voice its grievances."[13] The community had failed to establish collective bargaining, and it had also failed to generate "a spirit of human brotherhood," of "cooperation and genuine goodwill in all social relations," which alone could prevent such indignities. A "faulty social order" was at the root of the ghastly slayings and beatings. No other speaker so succinctly summarized the cooperative values that lay at the foundation of the New Deal, the labor movement, and the ethos of the 1930s.

Instead of channeling this moral indignation into a renewed drive for mass picketing or a general strike, however, the assembly at the opera house asked for investigations. The assault on the marchers was "a major breakdown of democratic government," and the resolutions carried that evening called on the US Senate's La Follette Committee and the federal government to take action. Investigating the connections between Republic Steel and the Chicago police would have little immediate impact on the steel strike. The "demand for an impartial investigation" might have carried more weight if it had enlisted the support of at least some of the Chicago's power brokers. As Professor Lovett noted, however, "no influential citizens came to our support, as I am sure would have been the case with the generation I had known in earlier days."[14] Lovett's comments overlooked the obvious limitations of the Popular Front. Steelworkers would have to fight for industrial democracy; middle-class liberals could offer support, but they tempered that with warnings of the purported imprudence of mass protests.[15]

No less remarkable is the fact that only weeks after SWOC's regional director, Van Bittner, dismissed the idea of a general strike, CIO members in Ohio launched one. In response to a court injunction that limited the number of pickets and prohibited interference with strikebreakers, CIO union leaders in Niles and Warren announced a general sympathy strike to support the steelworkers. Exhibiting remarkable unity, thousands of industrial workers and WPA employees joined the strike. Accompanied by several union wives, they marched through Warren; around 2,000 then assembled in front of that Republic Steel plant's main entrance. Some hurled epithets at strikebreakers leaving the mill; others turned over the automobiles of supervisors suspected of ferrying nonstrikers into the plant. Each group participating in the strike expressed the conviction that the version of law and order to which Chicago police chief Prendergast and SWOC leader Bittner subscribed could only work to the advantage of Little Steel. Yet in a conference with the local sheriff and the National Guard, SWOC leaders in Cleveland echoed those in Chicago's Southeast Side: collective protests threatened union respectability. The leadership in Cleveland would bring the general strike to an end in the belief that street demonstrations would compromise the union's appeal to a federal mediation board.[16]

Even as SWOC leaders made these decisions, United Auto Workers (UAW) union members at three plants in Pontiac, Michigan, announced a general strike in support of besieged Little Steel workers in Newton, Michigan. A massive contingency of UAW workers set out to reconstitute picket lines that the city's special police force had broken. Hopes for 15,000 to 20,000 additional autoworkers joining the strike dissolved, however, when UAW president Homer Martin threatened to expel workers who failed to respect their contracts.[17]

This grassroots activism was even more urgent now, since President Roosevelt had proven that he would not intervene. Despite appeals from the CIO, cabinet members, and the governor of Ohio for assistance, Roosevelt kept his distance. During a press conference with the governor, the president finally stated, rather cagily, "Common sense dictates that if a fellow is willing to make an agreement verbally, why shouldn't he put his name to it?" Bowing to pressure from Governor Martin L. Davey, however, Roosevelt established a federal mediation board to arbitrate the dispute and invited the companies and the union to the table. Republic Steel's president, Tom Girdler, made it clear that he did not respect the board's authority to settle the dispute. He would not sign an agreement; for that matter, he would not enter the same room as John L. Lewis and Philip Murray. Mediation failed to resolve the dispute, but Roosevelt also failed to assist the steelworkers who had voted for him, cheered him, and depended on him. At a news conference on June 29, the president threw up his hands. "The majority of people are saying just one thing," he told the group of attentive reporters, "'a plague on both your houses.'" Somewhere in the House of Labor a veil was torn. When John L. Lewis heard the news, he was seated on the edge of his desk, surrounded by reporters. "He said nothing," observed one newsman, "but his heels drummed against the desk's lower panels with a violence that just missed reducing them to splinters."[18] Lewis had learned a hard lesson on the limits of political loyalty. The workers would have to be the ones to determine the outcome of the strike.

Despite the Roosevelt administration's indifference to the Memorial Day Massacre, the steelworkers' unionization drive did not simply evaporate. Workers staged general strikes in two Ohio steel towns. In Chicago, picket lines held, albeit tenuously. Once again, Women's Auxiliary members played a key role. They persistently advocated for mass picketing. They also continued to sponsor "women's days" on the picket line *after* the Memorial Day Massacre. On June 4, the Women of Steel held a mass meeting to "hear speakers from the strike front at Republic Steel." On June 26, women took the lead in coordinating a "children's picket day" in Indiana Harbor. The daughters and wives and committed activists who had been so central to the organizing drive now sustained the movement at its critical point.

While Murray and other SWOC leaders appealed to Washington, DC, for relief, steelworkers turned to the support system they had built since the unionization drive began. On June 25, CIO workers held a mass meeting in Indiana Harbor that drew between 1,200 to 1,500 people, and they screened the Memorial Day newsreel to a dumbfounded audience. According to a police observer, a stockyard organizer by the name of McCarthy claimed that "the steel strike is the critical issue of C.I.O. and must be won at all costs." Intriguingly, McCarthy also claimed that had the strike been delayed by two months, meatpacking-plant workers would have had the union strength to "compel the steel companies to sign with C.I.O." In the meantime, he stated that the packinghouse workers would send pickets to the steel mill and would boycott the *Chicago Daily Tribune*. While Bittner assured authorities that there would be no "trouble" and Philip Murray urged the workers at US Steel to respect their contracts,[19] militant labor unionists called for action. Police observers reported that "Jesse Reese (colored) from Youngstown Sheet and Tube told crowd that if necessary they should die to win. He wanted more pickets on the line. Said it caused bosses to fear. . . . Forceful speaker well received." Strike leader and communist Joe Weber supported Reese. The steelworkers should be "willing to die for the cause and that the bloodshed would be on the heads of the owners of the steel mills."[20] Rank-and-file commitment had certainly not dissipated in Indiana Harbor.

What SWOC president Phillip Murray was never able to understand was that the steelworkers had transformed a labor dispute into a larger social movement. Those who believed that they had a stake in the fight—from sympathetic teachers to Women's Auxiliary members to packinghouse workers—elevated the contest above the limited issue of a signed contract. While many laborers viewed this as an opportunity to challenge the autocracy of the steel industry, Murray saw it as "abuse of our newly acquired power and influence" that would "ultimately destroy our union."[21]

Despite Murray's warnings to respect their contracts, progressive unionists at US Steel's Carnegie-Illinois plant tested the boundaries of SWOC policy. They attached their own simmering grievances to the Little Steel unionization drive. A broad cross-section of Chicago's working class believed that by confronting Girdler, they challenged industrial despotism and homegrown fascism. George Patterson and James Stewart, president of Local 65 South Works, channeled the moral indignation they felt after the Memorial Day episode into a resolution that condemned police brutality. They challenged the cautious legalism of SWOC leaders and demanded a measure of justice. "ECONOMIC ROYALISTS TO BLAME," the resolution declared, echoing how President Roosevelt had famously referred to the financial and industrial elites who opposed him during the 1936 election.[22]

Patterson's group went beyond condemnation, calling for a commitment to union democracy and social reform. "The issues are so much bigger than the strike in steel for a signed agreement, and they are no longer merely the CIO question, it is American Democracy that is being challenged."[23] The signers pledged to "do everything in our power to assist our striking steel brothers," since Girdler, Eugene Grace (the president of Bethlehem Steel), and the municipal officials who collaborated with them only wanted "to bring back the terrorized company steel town on a more definite fascist plan." In contrast, Patterson and his comrades committed themselves to a "democratic union," one in which none would be excluded on the basis of race, color, nationality, or "political belief." They wanted to carry on the work of educating the public about the "real issues of this brutal, unnecessary strike struggle" and the threat of "fascist tendencies" in America. Those same fascist forces had cried "'Communist' as they pull[ed] the trigger that sent striking steel workers to their deaths on bloody Sunday." Patterson's group had identified the larger significance of the Little Steel Strike. "We are not going backwards to company controlled towns," they asserted, "we are going forward with progressive, democratic America."

What was clear was that Patterson, Stewart, and the militant faction at Local 65 wanted to maintain the political character of the strike. They did not want it to degenerate into a dispute between one group of workers and their particular employer, which was precisely what Bittner seemed to favor. Instead, Patterson and his allies reminded Chicago's industrial laborers that this was a moment of *class* conflict, a confrontation between *all* workers and management over the basic issues of economic democracy and workers' rights.[24] It was the unwillingness of Republic Steel to engage in collective bargaining and the failure of Chicago's authorities to enforce the laws protecting labor unionism that had politicized so many workers in the first place. Even if the activism of the 1930s created the template, it was the actions of Tom Girdler and the police on the ground that catalyzed a broad panoply of the steelworkers' supporters and convinced them that mass picketing was necessary in order to win. To accept Bittner's injunction against "trouble" meant rejecting the political meaning of the strike. It also meant denying the reasons why so many had arrived at Sam's Place on Memorial Day.

Yet even Patterson and his militants drew back from direct action against Little Steel. They would respect the requirements of their collective agreement. It was certainly understandable why even Patterson's group was unwilling to risk a larger struggle: they had just won their own contract with management, the Depression was by no means over, the federal government was lukewarm toward unions, and middle-class opinion was turning against labor militancy. Yet in the formative moments of unionism in the steel industry, the workers' unwillingness to confront the leadership of their own union, let alone the nation's elites, would have enduring consequences for this movement.

Chicago Politics after the Police Riot

Public pressure did have an impact, since Mayor Kelly ordered the removal of nonstriking workers from the Republic Steel plant in Chicago's Southeast Side. But why had he been so indifferent during the Little Steel Strike? Why did he do little more than offer a perfunctory reassurance that "peaceful picketing" was protected by law? To understand this, we have to consider the political alignments in the city and the links between Chicago and Washington, DC, during the New Deal years. Elected in 1933, Edward J. Kelly inherited the political machine that his predecessor, Anton Cermak, had built so successfully. At the same time that Kelly strengthened it in Chicago, however, urban machines throughout much of the country atrophied.

Yet Kelly also inherited a city paralyzed by the Great Depression. Assisted by ward alderman and Democratic Party leader Patrick A. Nash, the mayor not only would improve the city's fortunes during this bleak period, but would also strengthen the alliances between Chicago's working class and the New Deal coalition. Confronted by a massive property owners' tax strike and protests at local banks by some 14,000 schoolteachers who had not been paid for months, Kelly authorized warrants on $1.7 million in unpaid taxes. He also spearheaded the movement for legislation that would permit the city to collect delinquent rents and income taxes. The revenue from these measures allowed the city to pay the teachers. At the same time, Kelly reduced Chicago's deficit and slashed the educational system's budget by closing schools, reducing programs, and laying off 1,300 teachers. He then justified the cuts as an expedient to save the system. An effective political operator, Kelly cushioned the opposition to his austerity measures by paying the salaries owed to municipal workers, who had also been working for free for months. Kelly's efforts at budget cutting, coupled with his delivery of valuable services to downtown merchants and his opposition to personal property taxes and state income taxes, won him the support of business owners. But his decision split Chicago's conservative middle class, who were outraged by the reductions in the public education system, from the liberals and reformers who supported the mayor's action. At the same time, Kelly won the support of the police and firefighters, who had been laid off or who had gone without pay in the early years of the economic crisis.[25]

If balancing the budget and reducing public expenditures was one factor in Kelly's success, the other was building up Chicago's infrastructure and providing for those hurt by the Depression. Both of these accomplishments flowed from his support for Roosevelt. By mobilizing the vote for Roosevelt, advocating for New Deal programs in the state, and defending the president's court-packing scheme of adding more justices to the U.S. Supreme Court, Mayor Kelly was able to channel substantial federal dollars into the Windy City. Through New Deal programs (such as the Works Progress Administration), Kelly built a new subway on State Street, upgraded the airport, expanded Lake Shore Drive, added miles to the city's parklands, and made substantial improvements to the city's road and sewer networks, all the while putting unemployed Chicagoans to work. Even though the relief payments and works projects came primarily at federal expense, Kelly was able to portray himself as the purveyor of public benevolence. He augmented this image by conspicuously reaching out to the city's African American community. Simultaneously, he rewarded his political subordinates for their support in getting the vote out. Kelly was Roosevelt's key agent in Chicago, and it was a relationship that paid enormous political dividends for the mayor.

The reciprocal relationship that Kelly established between city hall and

Washington, DC, influenced the mayor's response to the Little Steel Strike. When President Roosevelt distanced himself from the direct-action tactics of the CIO and, in effect, adopted a neutral position during the steel strike, Mayor Kelly did the same. He expressed perfunctory support for the CIO's right to peaceful picketing, but he maintained his allegiances to Chicago's police department and the Roosevelt administration. Mired in the court-packing scandal and damaged by conservative opposition to the wave of sit-down strikes, Roosevelt was content to let Little Steel and the CIO fight it out on the picket lines. Kelly saw no reason to object, particularly since, as historian Robert Slayton has observed, the Chicago machine had already forged ties to the conservative American Federation of Labor, which looked askance at the breakaway, "radical" CIO. Equally important, as Slayton has noted, the Chicago machine depended on major industries for campaign contributions, as well as other operating funds that were acquired through graft and kickbacks. The SWOC uprising threatened to disrupt this web of interdependent relations, which had already embraced the conservative trade unions and adapted to the challenges of the Great Depression by accommodating many of the pressures from below. Direct action by industrial workers challenged this comfortable arrangement.

Yet the repercussions that followed the Memorial Day Massacre forced Kelly to reassess his political calculations. In the wake of the May 30 incident, he would reach out to the CIO. He would even go so far as to recruit Harry Harper, the steelworker who lost an eye during the march, on his campaign team. By restraining the police department, conciliating the CIO, and offering concrete rewards for union support, Kelly and his successors incorporated the steelworkers' union into the political machine that dominated Chicago politics.[26]

That was what transpired in the long run, during the postwar era. In the short term, the steelworkers' strike faltered when SWOC leaders failed to capitalize on the indignation that followed from the Memorial Day Massacre. The powerful demonstration during the June 18 meeting at Chicago Stadium concluded with one question left hanging in the air: what should be done? Earlier, SWOC's Bittner had asked, "If we can lick the United States Steel Corporation, why can't we lick the Chicago police?" Yet the June 18 meeting ended without any resolutions for a general strike, a mass demonstration, or a series of boycotts.[27] The steelworkers would have to hope that the country's 600,000 mineworkers would drop their tools and join the strike. Since the city had forced Republic Steel to remove the non-striking workers from its Chicago plant, the steelworkers had an opportunity to shift the strike's balance of power in their favor. If the coal miners could launch a massive walkout, it might be enough to bring Little Steel to the bargaining table.

A larger, more coordinated movement built on worker solidarity throughout the entire industrial core of the nation might have turned the tide. Directly following the Memorial Day incident, police and public officials in Chicago held a meeting with union leaders at the Southmoor Hotel. When Captain Prendergast warned against any additional "trouble" at Republic Steel, SWOC's regional director, Van Bittner, complied. In fact, Bittner went so far as to declare that "there will be no more trouble with the Republic Steel Co. in Chicago."[28] On Tuesday night, June 1, as police mobilized for a rumored attack by Indiana workers supposedly armed with dynamite and shotguns, Bittner called the worried police chief to reassure him. According to Prendergast, Bittner reiterated his promise: "You can dismiss your police department. There will be no trouble." Prendergast agreed, and the shadowy saboteurs never materialized.

Yet Bittner also announced that there would be no general strike in Chicago. While Morris Childs, the secretary of the Communist Party in Illinois, called for a city-wide protest, Bittner resisted the idea of a larger rebellion. "A strike by the garment workers won't help the steel workers," the regional director claimed. He also restated his pledge to Prendergast. "I have given orders to the strikers that they be orderly, and I have told Capt. Prendergast that if they attempt to provoke any violence he should notify me, and I will put an immediate stop to it." Bittner, like SWOC president Philip Murray, was determined to reassure the public that the CIO was a legitimate and respectable organization. "We'll get those workers out of the plant without violence," Bittner told the press. "Violence never did anyone any good, and I'm not in favor of it." Yet by rejecting the idea of a general strike, insisting that the strikers behave in an "orderly" fashion, and asserting that he would control a supposedly frenzied rabble, Bittner implicitly validated the suspicion that the Memorial Day demonstrators had caused the "trouble." He unwittingly confirmed some of the worst antilabor stereotypes of the era. Moreover, he fuelled the communist-conspiracy theories disseminated by some of Local 1033's own members. By failing to distinguish between "violence" and legitimate mass protest, Bittner was indirectly rejecting collective action as a tactic for winning the strike. That shifted the locus of power from Chicago's

Southeast Side to Washington, DC, the president, and the courts. None of these had much interest in salvaging the steelworkers' cause.[29]

Vindication of a Kind: The La Follette Hearings

Convened on June 30, 1937, the La Follette Civil Liberties Committee hearings, led by liberal U.S. Senator Robert La Follette Jr. of Wisconsin, would expose the contradictions inherent in the official police version of the Memorial Day Massacre. The committee had already investigated the violation of basic civil liberties in the steel industry, the auto industry, and other notoriously anti-union businesses, which gave the labor movement an enormous morale boost and an additional measure of credibility. Now, the committee would reveal a pattern of complicity between the police and the steel companies which produced an inflexible intolerance of dissent. More than likely, Republic Steel supplied the police with tear gas. Approximately 40 of the officers carried special, white clubs that resembled hatchet handles. In their testimony before the La Follette Committee, Captain Thomas Kilroy would suggest and reporter Ralph Beck confirm that the company had issued these menacing weapons to the police. Moreover, Police Commissioner James Allman confessed that he might have been mistaken in his assumptions about the situation. His belief that the May 30 marchers had some premeditated, generally approved plan to breach the plant's gates and remove all of the "finks" who were still inside the steel mill was based solely on unreliable police informants.[30]

How, then, did Captain James Mooney determine the marchers' objectives? A friendly tip from a newspaper reporter, the captain claimed. Not a single independent reporter or observer could corroborate Mooney's assertion that those who assembled at Sam's Place under the hot sun of an early Chicago summer threatened to invade the mill. For that matter, Mooney's chilling account of a striker threatening to drive a meat hook "through your skull" lacked any substance: no witnesses, no threats, no meat hook.[31] Mooney's assertion that police had been firing in self-protection against fanatical union attackers was another fabrication. How could he square that claim with the fact that seven of the marchers had been killed by gunshots piercing their backs, and three from wounds in their sides? It strained common sense to suggest that an officer encircled by a menacing mob would be able to put a bullet in an attacker's back. "You can take the poorest marksman in the world," an exasperated Senator La Follette observed, "give him a gun, surround him with a bunch of men who are attacking him with clubs and let him shoot, and if he hits anybody I cannot see how he would hit him anywhere except in the front." The police officers' shocking accounts of a militant mob launching a frontal assault on the bluecoats, which the *Chicago Daily Tribune* had so smugly reported, did not stand up to scrutiny.[32]

It wasn't just the inconsistencies of the police officers' testimony that undermined their credibility; it was their callous behavior toward the wounded demonstrators.[33] Officer George Higgins typified the police brutality of that day.[34] He was the very same officer who claimed to have seen Joseph Rothmund carrying a nickel-plated revolver, a striker carrying a slingshot, Lupe Marshall carrying a bag of pepper, and a marcher wielding a sawed-off shotgun. He also asserted that he had heard the strikers announce, "We are going in that mill and drive them out." From Higgins's point of view, marching automatically made one a subversive. Under oath, La Follette Committee investigator Charles Kramer testified to the account Higgins had given of coming to the rescue of officer Walter Oakes, who was allegedly being attacked by one of the protestors: "I wait for my chance and measure him [the marcher] off, and, sock, I smacked him. . . . Oakes then got up and shot this Rothmund, . . . Oakes shot him again and perforated him in the stomach. This was the lousy communist." It was Higgins who remembered "the Porto Rican woman," Lupe Marshall, as a "communist." "I shoved her on her feet and she went down," Higgins told investigators. "I didn't hit her." It was Higgins and his crew that had prevented the drivers of the makeshift ambulances dispatched from Sam's Place from getting through to the wounded. "We let him come down and he asks, just like that, if anybody is injured," Higgins recalled in his interview with Kramer, "and I said, 'Get out of here you goddamned rat.'" Reflecting what would become the dominant opinion about police conduct that day, the *Washington Post* observed that the officers "have said nothing to contradict the general impression, drawn from sworn statements and photographs, that their men went berserk in turning back a platoon of strikers."[35]

The complexity of motives behind the conduct of Chicago's police force on Memorial Day requires some attention. Their tradition of strikebreaking antiunionism was critical. So, too, was the department's notoriety for indiscriminate violence and general unprofessionalism. Yet the attitude of the police toward radical dissidents, particularly those who belonged to ethnic minorities, compounded these factors. Captain James Mooney had developed his antipathy for communist agitators during the unemployment demonstrations of the early 1930s. In Mooney's opinion, Chicago's communists—marching boisterously, interfering in tenant evictions, and taking to the streets to demonstrate for strange and exotic international causes—had become "absolutely antipolice and antigovernment."[36] Traumatized by his memories of militant protests in the early 1930s, Mooney didn't view the Memorial Day events as a Popular Front celebration. Instead, he saw wild-eyed radicals who threatened his officers and wanted to break through the plant gates, attack the scabs, and subvert the established social order. "I will tell you my definition of a 'Red,'" said Mooney, lecturing Senator Elbert D. Thomas of Utah, the La Follette Committee's cochair. "He is here to undermine this Government and assault police. I don't know whether this is a good definition or not, but that is what I think. I am an American. My grandparents were born here." According to Mooney, the Memorial Day conflict was between Moscow-led radicals and the defenders of the American way of life. Mooney wanted to make a recommendation to the committee: "Deport every one of those Communists and all of those 'Reds' out of the country and then we will get along, they won't be assaulting policemen and dynamiting buildings, and then we will have a good Nation."

Mooney could not have identified a single communist convicted of dynamiting a building during the Great Depression, but his statement was not completely erroneous. Mooney was probably recollecting the bombing incident of 1910, when labor activists John and James MacNamara used dynamite to set off an explosion at the *Los Angeles Times*, a virulently antiunion newspaper. Samuel Gompers and the labor movement quickly came to the defense of the brothers, both members of the Iron Workers Union, which at the time was engaged in a tooth-and-claw struggle to defend the union shop in San Francisco. The explosion and subsequent fire killed 21 people. When the MacNamaras pleaded guilty, they dealt a blow to the labor movement from which it had yet to recover by the 1930s. Neither of the MacNamaras was a communist, or even a political radical, yet they had demonstrated their willingness to engage in violent extremism. Whatever distinctions may have been made between the MacNamaras' actions and political groups at the time, they meant little during the Great Depression. Opponents of the labor movement routinely conflated anarchism, communism, and labor violence. It was far more effective that way.

Many on the police force shared this inflexible anticommunism. Seventeen-year veteran Lawrence Lyons described the marchers as "the army of the 'Reds'" and believed that the "radical movement" had "brought these poor people to their destruction." Officer Jacob Woods concurred. "In the district where I worked, Senator, we are bothered with all kinds of people, with 'Reds.'" Recounting how Officer Walter Oakes shot and killed Joseph Rothmund, Officer George Higgins described the slain WPA worker as "the lousy communist."37 Following the massacre, police searched Lupe Marshall and Harry Harper for "communistic literature"; grilled demonstrator Max Guzman about alleged communist connections; and arrested George Patterson, Louis Selenik, and Joe Weber in the Chicago force's hunt for red subversives. Two weeks after the Memorial Day incident, Lieutenant Make Mills, the Red Squad leader in charge of industrial surveillance, reported to Police Commissioner Allman that he had identified 13 alleged communists in the steel movement.38

Police testimony exhibited more than hysterical anticommunism; it also revealed the level of ethnic and racial bigotry on the force. "The class of people that live and work in these mills," Officer Lyons announced, "are of foreign extraction—I mean are foreigners to this country . . . where they have talked the foreign language in their home for years and probably haven't got as much respect for the American flag as I have."39 Patrolman George Higgins echoed Lyons's sentiments. Higgins had been "in that race riot"—the 1919 pogrom that left 23 African Americans and 15 whites dead. He had also been in the stockyards strike, "with all them foreign savages out there," but the Memorial Day incident was of another order entirely.40 When Captain Mooney explained to Senator Thomas that communists only wanted to attack the police and subvert the government, he added that "I am an American. My grandparents were born here."41 This xenophobic strain blended with the officers' contempt for "agitators" and reinforced the stereotype of foreign subversives leading liberty-loving Americans into peril.

By focusing its laserlike attention on police misconduct, the La Follette Committee inadvertently shifted attention away the political culpability of Mayor Kelly and the Roosevelt administration. The workers and their supporters, inadequately protected in the exercise of their rights, waded into a disaster waiting to happen. That bloodbath was only compounded by the subsequent unwillingness of the federal government to enforce labor rights and compel Little Steel to bargain collectively. Republic Steel and the Chicago Police Department bore the principal responsibility for the tragedy of Memorial Day 1937. But by adopting a hands-off policy designed to placate conservative middle-class opinion, state and federal authorities created the conditions for the incident in Chicago's Southeast Side. Their stance, together with the timid leadership of SWOC and the reluctance of so many rank-and-file steelworkers to question those leaders, ensured the subsequent defeat of the Little Steel Strike.

What is easy to lose sight of is how the Memorial Day Massacre fit into a larger pattern and a longer tradition of antilabor violence. Seasoned labor journalist Mary Heaton Vorse, a close observer of the steelworkers' strike in Ohio, reminded her readers of its longer historical context. Writing in 1938, Vorse concluded that utilizing the police to deny workers their constitutional rights had become a repellant American tradition. "The shooting of workers in steel began in Homestead in 1892 and has gone on steadily ever since." Vorse recited a litany of casualties from the history of American labor: 21 killed in the steel strike of 1919, including labor activist Fannie Sellins; "uncounted" victims slaughtered in the efforts of the United Mine Workers to organize; 19 victims dead in the Ludlow Massacre; 7 textile workers and the local sheriff killed in Marion and Gastonia, North Carolina, in 1929; and 7 workers shot in the back at Honea Path in South Carolina during the great uprising of 1934. Her enumeration left out 123 workers and militia killed during the unprecedented conflict known as the Great Railroad Strike of 1877. Nor did Vorse mention the 14 who lost their lives during the anthracite coal strike of 1902; the 13 workers shot and killed at Mckees Rock, Pennsylvania, during a steel strike in 1909; and the 4 marchers gunned down during an unemployed march on the Ford Motor Company's River Rouge plant in Dearborn, Michigan, in 1932. Even that accounting does not exhaust the carnage of America's labor wars, or the violent clashes of the Great Depression years. Yet Vorse had made her point: the constant purpose of state-sanctioned and company-sponsored violence was not to protect law and order, but to undermine the effort of workers to organize.

In the case of the 1937 steel strike, however, violent reprisals had an added dimension. Vorse wanted her readers to know about the tradition of antilabor violence in America, but she also intended to explain the political calculations underlying the virulent opposition to SWOC in the months prior to the strike. Those tactics expressed a more ambitious strategy: to undermine the movement for social democratic reform, and overturn the New Deal itself. "The steel strike was not the usual industrial conflict," Vorse reflected. "It was a challenge to the New Deal in which it was partially successful, since the independent steel companies literally got away with murder and succeeded in defying the National Labor Relations Act." Roosevelt's New Deal program had already suffered a major setback when Congress and many of the president's own supporters rejected Roosevelt's bid to pack the Supreme Court with justices favorable to the New Deal. Now it seemed that the drive to give meaning to the Wagner Act, one of the few New Deal measures upheld by the Supreme Court, was in jeopardy. The antiunion campaign in Chicago was directed at more than stopping SWOC in the Little Steel Strike; it was aimed at disrupting Roosevelt's coalition by attacking its largest and most vocal constituent: the CIO.[42] The Roosevelt administration aided and abetted this effort by abandoning the steel movement at the most crucial moment.

By 1937, the political tides had shifted against the New Deal, and the union movement in Little Steel was caught in a dangerous undertow that threatened the whole concept of a more progressive New Deal. The liberal-labor alliance confronted an emerging coalition between conservative southern Democrats and Republicans determined to obstruct the administration's wages and hours bill. At the same time, it encountered a congressional movement to investigate alleged subversives in the federal government. Those reinforcing Little Steel included the right-wing Liberty League and the U.S. Chamber of Commerce, as well as "the other powerful anti-Administration forces in Congress and outside."[43] What was at stake in the Little Steel Strike was not only the possibility for democratic unionism at the ground level, but the future of an expanded vision of social security and economic democracy.

Despite the scandalous revelations during the La Follette Committee hearings, the police version of the Memorial Day events struck a chord. That was because journalists such as Westbrook Pegler, and newspapers like the *Chicago Daily Tribune*, transformed the men in blue into heroic defenders of law and order. In some sense, the police had joined private antilabor detective firms, such as the Pinkertons, to become the paramilitary wing of the anti–New Deal movement. The strike that began as a dispute over union recognition had escalated into a struggle over the entire idea that collective, working-class representation had a legitimate place in American life.[44] If organized labor was respectable, patient, and moderate, then it might be tolerable. After all, US Steel had finally agreed to permit it. Yet if organized labor engaged in direct-action protest as part of a larger movement for social reform that might challenge managerial prerogatives and company profit margins, the boot, if not the bullets, of state coercion might be justified.

If most American elites refused to speak out against this rising tide of anticommunism, U.S. Representative Maury Maverick of Texas certainly would. Congress had been echoing the tensions generated by the strike for weeks. Media accusations of CIO "irresponsibility" and communist infiltration gained a hearing in Congress, as opponents of a progressive New Deal showed their colors. When the steelworkers' strike escalated in Ohio and the La Follette hearings came to an end, the intensity of the friction between the opposing factions became white hot. U.S. Representative Eugene Cox of Georgia castigated the CIO as "hysterical, highly provocative, and calculated to bring bloodshed and disorder." Cox and a colleague, Republican Congressman Clare E. Hoffman, had also volunteered to lead a paramilitary group of "patriotic citizens" to protect Monroe, Michigan, against the supposed attacks of CIO picketers. Cox's inflammatory remarks about violent southern resistance to any CIO campaign in that region provoked a response from Maverick that exposed the motives underlying the anticommunist united front. "Who is calling for blood and violence?" Maverick asked in a speech in the U.S. House of Representatives on July 2, the final day of the La Follette Committee hearings.[45] "Why, gentlemen, those who prate about the preservation of the Constitution, those who wear patriotism on their sleeves, those who call themselves conservatives and wrap themselves in the flag." Yet none of these "conservative gentlemen" had denounced the murder of 10 marchers exercising their constitutional rights in Chicago's Southeast Side. The Memorial Day Massacre was "one of the bloodiest, most shameful pages in our history," Maverick passionately declared, but "no leading conservative denounced it." He also offered an historical parallel that sounded an ominous warning of what might be the result of moral indifference to the labor movement. The American Revolution, Maverick declared, grew from the intransigence of conservatives "too stupid to see they were forcing the Revolution" on Americans.

Maverick castigated the "patriotic talkers" who refused to acknowledge the wholesale violation of workers' civil liberties, but instead denounced communism. "Oh, my colleagues, the old cry of communism is getting very thin. It gets thinner and thinner, answers no arguments, reveals no facts, settles no problems. John L. Lewis! Civil War! Communism! Communism! The

Red Flag of Russia!" But the labor movement was not simply Lewis or the CIO, Maverick argued. "It is a movement of the American people." Maverick could find no communists in the CIO, which suggested he wasn't looking closely enough. Yet he understood that the dominating spirit of the movement was the drive for basic civil liberties and industrial democracy. "The working man of America is not a Communist, he is not a coward, and he is not a sheep." Maverick denounced the ideological and actual violence that his congressional colleagues threatened and called on them to respect the provisions of the National Labor Relations Act.[46]

Despite this defense of labor rights, the arguments against progressive labor unionism gained credibility under the mantle of anticommunism. Political leaders opposed to the New Deal, industrial executives hostile to workers' rights, and media leaders antagonistic to social reform formed a united front, and these forces would escalate the attack on the New Deal. They simultaneously derailed the drive for industrial unionism and established a new antireform coalition in American politics. Maury Maverick would be one of its first victims. The following year, he went down to defeat as conservatives trumped several of the president's liberal candidates. In summer 1937, the conservative coalition actively sought to undermine the industrial union movement of the 1930s.[47]

A Shift in Public Opinion

The report that the Senate Subcommittee on Education and Labor issued on July 22 almost completely vindicated the demonstrators. Some of the marchers probably hurled profanity at the police and waved their arms excitedly to make their case for picketing in front of the Burley Avenue Republic Steel plant. "But there is no evidence of physical threats," the report stated, "or of the frenzied disorder which the police describe."[48] Carefully weighing the testimony of reporter Ralph Beck and Captain James Mooney, the subcommittee concluded that policemen had fired the first three shots before the strikers had launched any missile barrage. Focusing more on the stick that Beck saw flying through the air than the "backward movement" that Frank McCulloch observed, subcommittee members determined that "the first shots came from the police; that these were unprovoked, except, perhaps, by a tree branch thrown by the strikers, and that the second volley of police shots was simultaneous with the missiles thrown by the strikers." In addition to what was captured on the Paramount newsreel, the photographic evidence from still cameras confirmed that the left side of the strikers' line was in full retreat well before the first devastating volley was fired by the police.

What the subcommittee did not seem to consider was the possibility that one or more of the supposed warning shots fired by nervous police officers may have triggered the fusillade. Since Mooney's verbal commands had been inaudible to those standing more than a few feet away from him, most officers probably would not have been able to determine *why* those shots had been fired or whether there had even been any warning. Taking into account the tense circumstances, the heat of a late Sunday afternoon, the policemen's limited training in crowd control, the officers' inadequate command and control on the field, the simmering antagonisms built up over the previous week, and the fact that shots had already been fired during an earlier confrontation, it was not a surprise to see the police using their weapons once again, this time to deadly effect. That escalation into violence had certainly been a common pattern for labor disputes in the 1930s.

The subcommittee's investigators, systematically dissecting each moment in the Memorial Day episode, discredited the idea that a fanatical mob attacked a stoic line of disciplined policemen. In addressing the behavior of the police following the assault, Senators Thomas and La Follette could barely contain their indignation. The police made no effort to render first aid on the field. They did not try to distinguish between the seriously and superficially wounded. Captains Mooney and Kilroy did nothing to coordinate treatment for the wounded. Instead, officers indiscriminately hurled seriously wounded strikers into waiting paddy wagons. Analyzing the photographs taken that day, Thomas and La Follette concluded that "the police dragged seriously wounded, unconscious men along the ground with no more care than would be employed on a common drunkard."[49] Nothing was more egregious than the treatment of Earl Handley, bleeding from lacerations to an artery and a vein in his leg, yet hauled out of the union vehicle carrying him to safety. Like Kenneth Reed, the striker who died en route to the hospital in Lupe Marshall's lap, Handley was tossed indifferently into one of the overcrowded paddy wagons.

The evidence overwhelmingly appeared to support the conclusion that the aggressive police response was unjustified and inexcusable.[50] The subcommittee judged that the use of excessive force "must be ascribed either to gross inefficiency in the performance of police duty or a deliberate effort to intimidate the strikers." After all of the appeals by Bittner and the Steel Workers Organizing Committee to address what was at stake in the Chicago incident, elected officials—Thomas, La Follette, and their subcommittee—finally stated what had seemed so obvious to union members: the police had operated as strikebreakers. After discovering that the Republic Steel Company had supplied, fed, and harbored police officers, it would have required willful ignorance to conclude otherwise.

In Chicago, the citizens' commission formed at the Civic Opera House protest on June 8 corroborated the findings of the La Follette hearings.[51] Moreover, Paramount released a modified version of the footage newsreel photographer Orlando Lippert recorded on Memorial Day, which triggered a discernible shift in opinion. On July 3, the film was screened for the first time in New York City. At the Paramount Theater in Times Square, the "violence of the action" on May 30 in Chicago "seemed to stun the audience."[52] The sight of police officers charging and firing at retreating strikers produced a "collective gasp" from the audience as it watched in "utter silence." Eyewitness testimony had vividly illustrated the extent of police brutality on Memorial Day, but the Paramount newsreel had a uniquely visceral impact on viewers. The wider public reaction mirrored the response that the film generated in the congested chambers on Capitol Hill where reporters, congressional representatives, and witness first viewed the notorious footage. There, Lippert recounted how demonstrators marched across the desolate field, coming face to face with the line of bluecoats. He had captured it all on film, with the exception of seven seconds, during which he changed lenses for closer shots. Yet it was during those seven seconds that the assault began. Nothing would prove to be more intriguing yet more frustrating than the possibility that Lippert might have been able to document the spark that started the attack. He claimed to have observed objects being thrown from the ranks of the strikers, but his statement was hardly a substitute for those seven seconds of missing film.[53]

The newsreel utterly decimated the police version of events. It documented the tense but nonviolent exchange that preceded the clash. It showed policemen carrying white hatchet handles, provided courtesy of Republic Steel. It revealed officers drawing and firing their weapons. It verified that the workers were in flight from the bluecoats' onslaught of tear gas, billy clubs, and gunfire. Instead of wild-eyed zealots, the film captured terrified men and women fleeing a torrent of swinging billy clubs. It showcased the relentless beating of fallen marchers and the vigorous pursuit of others. It clearly recorded the tangled mass of wounded and fallen marchers directly opposite the center of the police line. It showed the surprising number of women in attendance during the march, many of whom strode to the front of the column and endured the worst of the violence. Lupe Marshall was not the only female who was wounded that day; Ada Leder and Catherine Nelson each sustained a bullet wound to the leg. More than anything else, the film captured the inhumane treatment of the wounded and dying after the attack.[54]

The newsreel footage may not have solved who had started the clash, but it decisively exposed the police misconduct *following* the massive fusillade, an example of the legendary brutality of a police force seasoned by years of internal corruption and Wild West lawlessness. Lippert's camera unflinchingly documented policemen unceremoniously shoving wounded strikers into paddy wagons, with some of the marchers clutching bleeding scalps and others dizzied and disoriented. Alfred Causey is seen helplessly writhing and then dying in the dirt road before one police officer gently places a piece of cardboard under his head, the only gesture of compassion documented that day. Earl Handley can also be seen, semiconscious and mortally wounded, as police officers drag him to a paddy wagon. No incident in the film was more poignant than Otis Jones's feeble efforts to stand after being paralyzed by a gunshot to the spine. The June 17 issue of the *Washington Post* reprinted an article by journalist Paul Y. Anderson, who described the pathos of Jones's struggle: "He moves his head and arms, but his legs are limp. He raises his head like a turtle, and claws the ground."[55]

The film erased any doubt about who was responsible for the bloodshed on Memorial Day. Many of the mainstream daily newspapers now condemned the brutality of the officers and their men. Rather than defenders of law and order, the police looked more like jackbooted thugs. "Guns fired point blank at fleeing marchers," the *Washington Post* indignantly observed, "prone figures clubbed unmercifully as they groaned in death's agony on the ground, policemen gone berserk in a bloody holocaust of madness."[56] Anderson's report and the testimony in the Washington, DC, hearings convinced several major newspapers to present a more sympathetic treatment of the strikers. With the widespread dissemination of the film—the notable exception being in Chicago, where the police department's Movie Censorship Bureau banned it—several newspapers that had jumped in to condemn the marchers now rebuked the police for their evident brutality.[57] To supporters of the steelworkers, it seemed as though they had reached a critical turning point.

The Counterrevolution in Chicago and the Challenge to the New Deal

Despite SWOC's hopes, the coalminers never did walk out in support of the steelworkers. The general strike that Morris Childs of the Communist Party had called for never did happen. Most leading newspapers eventually condemned the police brutality of Memorial Day, but this failed to touch off a firestorm of indignation. Conservative newspapers, such as the *Chicago Daily Tribune*, and antiunion journalists like Westbrook Pegler stood firm. They tried to rehabilitate the image of the police as the protectors of respectability, property, and the status quo, instead of hooligans in uniform. How had all this come about?

For a variety of reasons, a powerful segment of American society had grown increasingly skeptical about the New Deal. What this coterie held in common was the view that the labor movement had grown too powerful. White segregationists were afraid that the New Deal's social programs would deprive the South of its immobile and cheap African American labor force. White southerners who owned the mills and cotton fields in the region feared unionization, because it would disrupt the biracial labor market that kept blacks and whites divided and compliant. Struggling small businesses joined bigger manufacturers in believing that they could not absorb the cost of paying higher wages in an intensely competitive economy. Southern industrial boosters—who believed that unionization would undermine the region's comparative advantages of cheap labor, minimal governmental intervention, and low taxes—supported them eagerly.[58]

In order to understand the broader repercussions of the Little Steel Strike, it's important to examine this southern wing of the anti–New Deal movement a little more closely. In the early months of 1937, Roosevelt had been preoccupied by his fight to expand the number of justices on the Supreme Court, in order to guarantee the constitutionality of key New Deal measures. Leading media outlets, a majority of congressional Republicans and several Democrats opposed the president's court-packing scheme, arguing that it was an exercise in dictatorial authority. Southern congressmen became its most vociferous critics. They feared that the appointment of liberal justices to a predominantly conservative court might threaten racial segregation in the South. Southerners had already made sure that the Social Security Act did not cover agricultural and domestic workers, most of whom were African American. Their resistance to the New Deal's progressive reforms only mounted when Roosevelt unveiled his plan for a wages and hours bill, the measure that would become the 1938 Fair Labor Standards Act. Southerners vigorously opposed a minimum wage that would jeopardize the region's low-wage advantage. For the same racial and economic reasons, they castigated the CIO and the labor militancy of the Little Steel Strike. Southern Democrats would join conservative Republicans in working to repeal progressive taxation and federal job-creation programs. In the postwar years, these two factions would combine forces to pass the Taft-Hartley Act, which would drastically limit the rights of workers to organize and challenge managerial authority.

Certainly the sit-down movement was at the center of this deepening class divide, but other incendiary elements contributed to the growing anti–New Deal reaction. The court-packing crisis was one of them, but the "Roosevelt recession" was another. As a result of a combination of Federal Reserve Board policies that increased interest rates, the impact of the first Social Security tax levy, and the president's decision to balance the budget through massive spending cuts in the WPA and the Public Works Administration, economic recovery stalled, threatening a return to the depression. Unemployment skyrocketed to 11 million as Americans began to question whether all of the New Deal's social engineering had achieved any lasting improvements. While working-class Americans turned against the Roosevelt administration for not having done enough, many middle- and upper-class Americans blamed it for having done too much.[59]

Even before this recession really took hold, wealthy and property-owning Americans questioned the nation's political direction. Reading about a wave of sit-down strikes, wildcat job actions, mass demonstrations, and violent picket-line confrontations—which the newspapers routinely blamed on strikers—many middle-class Americans came to the conclusion that organized labor had gained too much power. They believed that workers had abused the rights granted to them by the courts and the federal government. Yet the mainstream media consistently ignored the history of antilabor violence at the hands of private security forces, municipal police, and the National

Guard when they reported on the sit-down strikes. More importantly, the mainstream newspapers and radio stations failed to explain that workers and owners never really had been on a level playing field. When the steel companies systematically flouted national labor law, which had finally granted workers a measure of protection, the workers felt compelled to take direct action. What is important here is to remind ourselves of the context for their response: American business had viciously resisted the appeals for workers' rights in the past. In the midst of an economic crisis that left workers utterly bereft of economic security, action seemed all the more necessary. For most industrial laborers, the Roosevelt administration and the Wagner Act meant nothing less than freedom from fear.

Despite the bloodshed of the Memorial Day Massacre, the antagonistic view that much of the middle and property-holding classes had toward labor activism intensified during the Little Steel Strike. The distorted media coverage of the incident on Chicago's Southeast Side added fuel to the fire, but events in other areas of the nation affected by the strike added to the overall negative opinion of labor unionism. When striking workers interfered with the efforts of the Post Office to deliver the mail—which included covertly packaged supplies for nonstriking workers—to the Republic Steel plant in Warren, Ohio, the newspapers and congressional conservatives condemned the picketers' actions. Public opinion went from bad to worse when striking workers fired at planes that tried to conduct an airdrop to support the workers inside the plant. In Youngstown, Ohio, a conflagration broke out when company officials tried to drive through picket lines to deliver food to nonunion workers inside that Republic Steel mill. Police and strikers also engaged in a pitched, day-long battle that left one worker dead after authorities tried to break up women picketing at the plant. In Johnstown, Pennsylvania, unionists and nonstriking steelworkers engaged in hand-to-hand combat in the streets. Similar episodes occurred in Monroe, Michigan. All of these conflicts illustrate how divisive and bitter the Little Steel Strike had become. Yet conservative newspapers laid the blame for the confrontations squarely on the shoulders of John L. Lewis, the CIO, and the Roosevelt administration.[60]

Above all, the wave of strikes in 1937 had created an atmosphere in which middle-class conservatives would conclude that Chicago's Memorial Day incident was a regrettable but avoidable result of labor radicalism. Approximately 400,000 workers had launched sit-down strikes that year. Laundry workers, sales clerks, stock boys, waitresses, sanitation workers, and even

WPA employees took to sit-down tactics in a frenzy of union organizing.[61] What was exhilarating to American workers, however, was threatening to small-business owners, professionals, and financial elites. Many middle-class Americans resented what they considered to be the cavalier willingness of employees to violate the property rights of business owners and stockholders. The mainstream media encouraged this response, portraying the CIO as a maverick upstart that had moved too aggressively and acquired too much power.[62]

More than this, it seemed as though industrial unrest was increasing, despite the advent of the National Labor Relations Act. Instead of guaranteeing labor "peace," as the act's proponents in Congress had promised, labor antagonism seemed to be intensifying. For tactical reasons, some American businesses had submitted to labor unionism. Yet Tom Girdler probably spoke for more than Little Steel when he appeared before the Senate Post Office Committee to testify about strikers obstructing the delivery of supplies management mailed to nonstriking workers inside the Republic Steel plant in Warren, Ohio. In a heated exchange with Senators Joseph Guffey and Kenneth McKellar, Girdler exploded. "I'm trying to tell this committee I won't have a contract with an irresponsible, racketeering, communistic body like the C.I.O."[63] For many small and mid-sized businesses that operated on margins much thinner than those of Republic Steel, Girdler's words echoed loud and clear. The CIO had become a threat to *their* sense of security, not to mention their self-interest. The multiple sit-down strikes had vividly highlighted the nation's class divisions. They had driven many to take a position against the industrial labor rebellion. Middle-class Americans who believed that the wave of sit-down strikes had gone too far worried more about the increasing effectiveness of the CIO than about the bloodshed in Chicago on Memorial Day. Congressional conservatives, particularly those from the South, joined them.

It was these sentiments that ultimately mitigated the political impact of the La Follette Committee's shocking revelations.[64] According to the *Chicago Daily Tribune*, the committee had "repeatedly performed services of the highest value to the cause of insurrection in this country."[65] It had become an instrument of "C.I.O. propaganda" by endorsing the sit-down strikes and acting as a "publicity agent for the party of revolt." The Paramount newsreel and the revealing testimonies of Lupe Marshall, Harry Harper, and Dr. Lawrence Jacques had done nothing to revise the *Daily Tribune*'s earliest conclusion, which laid the blame for the Memorial Day Massacre squarely on the shoulders of the demonstrators. The *Daily Tribune* was not the only newspaper in Chicago, of course, nor could it claim to represent the views of the majority of Americans. Yet the paper did reflect the opinions of Chicago's power brokers, middle-class business owners, and property holders, and it is not too much of a stretch to suggest that it promulgated the views of an American elite who had always been suspicious of the New Deal. The Little Steel Strike became a galvanizing moment for the conservative opposition. Business leaders echoed the *Chicago Daily Tribune*'s opinion that radical agitators bore the ultimate responsibility for the Memorial Day Massacre. In an interview with *Fortune* magazine, a group of executives did acknowledge that authorities had used excessive force.[66] Even so, the magazine reported that the "consensus of round-table opinion was favorable to the police." One of these business leaders seemed to encapsulate the conclusion that middle-class Americans would settle on in the aftermath of the La Follette Committee's hearings. "The strikers went out there for trouble and they got it," he asserted. "The stanchness of the police has administered a healthy check to irresponsible elements." In essence, the police may have behaved badly, but the marchers had it coming.

5 Little Steel and Class Warfare

ONE OF THE GREATEST CHALLENGES facing the steelworkers since the beginning of the union organizing drive was the cautiousness of their own leaders. Philip Murray and SWOC spearheaded a massive strike against one the most powerful industries in America at precisely the moment when a large segment of middle-class opinion was turning against organized labor. The wave of sit-down strikes that dominated the headlines in 1937 had turned many Americans against labor unionism. While polling data suggested a tolerance for labor unions in 1936, opinions turned sharply against them after the labor uprisings of 1937.[1] White southerners, northern Republicans, and even some northern Democrats seemed to be losing patience with the CIO's militancy.

It was in this environment that Murray and other SWOC leaders sought to convince Americans of the law-abiding respectability of the CIO. It seemed particularly urgent to do so at this time, since SWOC was fighting the Little Steel Strike very shortly after it had won a major breakthrough at US Steel. That victory in itself illustrated the worldview of the union's upper echelon, which privileged high-level bargains won by charismatic leaders over grassroots activism and shop-floor democracy. This helps to explain why John Riffe stifled the sit-down strike at Republic Steel's Burley Avenue plant, why Van Bittner insisted that the steelworkers would not "cause trouble," why John L. Lewis spent most of his time in Washington, DC, and why Philip Murray insisted that the employees at US Steel's Carnegie-Illinois South Works mill "honor their contracts" rather than walk in support of the Little Steel workers. These top union men worried not only that police violence would be blamed on the steelworkers, further souring public opinion toward the labor movement, but also that a violent, protracted labor struggle would damage the Roosevelt administration, the union's lifeline to legal legitimacy and political influence.

Nonetheless, what seemed like sound pragmatism at the moment further eroded the possibility of winning the Little Steel Strike and advancing the social movement of the 1930s. One of the central reasons for this failure was the wall of media hostility that confronted the steelworkers in 1937. Little that Philip Murray or Van Bittner could say in the mainstream media would convince middle-class readers of the respectability of the CIO after months of negative newspaper editorializing. Bittner and Murray wanted to assure the public that the CIO would honor its contracts, play by the rules, and promote industrial harmony. Yet SWOC's leaders were making their case at a time when Little Steel was waging open war against industrial unionism. Their call for respectability was the equivalent of one-way disarmament. By insisting on federal mediation and contract compliance, Murray was depriving workers of the weapons they needed to challenge the antilabor, antiprogressive alliance that was gaining strength in 1937. Short of a major shift in media reporting, the assurances of these SWOC leaders could do little to persuade public opinion of the justice of labor's tactics. They had certainly failed to convince Roosevelt.

Writing to the editor of *Nation* magazine, Rabbi Harry J. Brevis of Buffalo, New York, summarized the forces that SWOC was up against. On Friday, June 18, Brevis listened to seven national and local radio programs dedicated to a consideration of the strike. "Not one of the speakers had a good word to say for the C.I.O. and John Lewis or a word of criticism of the steel barons."[2] Brevis noticed no distinction in the tone or substance of the discussion between fiery antilabor demagogues such as Gerald L. K. Smith and the "high-price and presumably 'impartial' news commentators" weighing in on the momentous strike. Brevis, claiming a neutral position, questioned "whether such one-sided propaganda is fair to the thousands of C.I.O. members, to the millions of still unorganized workers who, sooner or later, will have to take a definite stand" on industrial unionism. Equally important, Brevis questioned whether the media's strident antilabor bias was fair to "the broad masses of American citizens whose aggregate opinion must ultimately play a part in the settlement of the controversy." Brevis's analysis effectively encapsulated the difficulties that any social protest movement would face in modern America.

Yet Brevis had also honed in on the crucial issue facing organized labor at that moment, a challenge that would persist into the twenty-first century. It could appease conservative middle-class opinion and hope for federal intervention in the future, or it could mobilize the working-class majority, push to win the strike—as had happened at the automobile plant in Flint, Michigan—and hope to swing the balance of power in its favor. Murray and SWOC chose the former. Observers during that period understood the implications of labor's excessive dependency on the federal government. Looking at the Ohio strike zone, where SWOC leaders applauded Governor Martin L. Davey for sending in the National Guard to supposedly restore order, the *Nation*'s editors saw disaster.[3] This state was not Michigan, where its governor, Frank L. Murphy, who had been elected on the strength of political support from labor, opposed the use of the National Guard as a strikebreaking force during the workers' sit-down in Flint. This was Ohio, the "apotheosis of the Middle Western middle class," where the governor mobilized the National Guard to break strikes and appease middle-class opinion. Contrary to CIO illusions, Ohio's governor did not dispatch the troops to protect striking workers. In 1937, SWOC was fighting steel companies that dominated the small towns in which they operated. CIO militancy had a much greater chance of succeeding in urban areas, where a critical mass of support for the union could affect the balance of power.

The *Nation*'s editors criticized the labor leaders' naïve attitude toward martial law, but it was the union's reliance on the Roosevelt administration that generated the greatest concern. According to one report, workers in Johnstown, Pennsylvania, had signed a petition calling for compulsory arbitration. This was symptomatic of a movement that is "counting far too much on the federal government, particularly on Roosevelt."[4] What SWOC was up against was not only the police and the National Guard, but "a section of industry intrenched [sic] in the natural social backwardness not only of American independent business and middle-class opinion but of great sections of American workers." Equally importantly, SWOC was confronting a Democratic Party that wanted to validate but also contain the labor movement. If the strikers had won, this would have meant rethinking the labor movement's relationship with the Democratic Party, and effectively transforming the Democrats into a party of labor. Considering the party's southern representation and Roosevelt's own pragmatism, that was a remote possibility.

According to the *Nation* magazine, what compounded the challenge was the fact that the CIO itself was "politically and socially backward." The CIO was failing to build a grassroots movement that could transform the way that workers understood the forces aligned against them. At the same time, a local movement that carried the potential for linking industrial unionism and social democracy was developing in Chicago. It produced what historian Ian McKay has called a moment of "supersedure," or collective understanding.[5] Through newsletters by the union and the Women's Auxiliary, CIO publications, sympathetic newspapers, and the outpourings of supportive writers, workers in the city's Southeast Side began to shape a narrative that made sense of their experience.

Yet even here an overreliance on the Roosevelt administration compromised the movement. At a critical moment, the president, through his famous "a plague of both your houses" statement, established a moral equivalency between striking workers and Republic Steel's president, Tom Girdler. This only legitimized the negative opinion of the labor movement then solidifying in Congress and the media. The steelworkers' support for a federal governmental solution might have been understandable during the Little Steel crisis, but the *Nation's* editors warned that "such precedents, forged in the heat of battle, may well cool into steel nets for restricting labor's rights, and with them the rights of citizens in general." By the 1940s, those nets had become very rigid indeed.

Why, then, did SWOC's leaders dismiss the idea of a general strike or a more widespread protest? In part, they were philosophically opposed to mass movements, at least those that they could not control. Murray and Lewis saw themselves as the executive branch of the labor movement: they did the negotiating, and they made the important decisions. Even as the organizational drive for unionization gained momentum, SWOC's upper echelon struggled to impose their conservative vision of unionism on the general membership. As early as January 1937, regional director Bittner was making it clear to SWOC organizers precisely who was in charge and what the objectives of the movement would be: "We are dictating policies of all lodges until steel is organized. . . . Democracy is important but at this time collective bargaining & higher wages is [sic] the issue."[6] Against the growing willingness of SWOC members to see the steelworkers' struggle as one part of a larger movement for social democracy, Bittner and Murray strove to narrow the movement's focus. "We are dealing with organizing steel workers & this only, forget about Spain situation, auto situation, & all other world problems." Yet this autocratic tendency ran counter to the egalitarian ferment at the local level, where steelworkers engaged in sit-down strikes, women staged protests in downtown Chicago, and African American and Mexican steelworkers tasted leadership roles in the front ranks. Since the beginning of the unionization movement, John L. Lewis and his allies had promised peace in the labor force in exchange for collective bargaining agreements. They abhorred the idea of "wildcat" strikes that undermined a union's trustworthiness and their own "respectability." Yet labor peace came at a very high price.

In the immediate term, SWOC's unwillingness to support a larger protest hastened the failure of the strike. John L. Lewis's decision to attend to federal mediation in Washington, DC, rather than address a mass meeting at Chicago Stadium symbolized SWOC's dependency on the national government. In the aftermath of the Little Steel Strike, SWOC president Phillip Murray would endorse "tight control and internal discipline" to preserve the flailing union. That was precisely the kind of leadership that regional officials Van Bittner and John Riffe had tried to impose on Chicago's Southeast Side.[7] Addressing a radio audience in 1939, SWOC district secretary Samuel C. Evett insisted that "the outstanding and basic principle of the C.I.O. and all of its affiliates is that of responsibility. It is fully realized and recognized that only through strict observance and enforcement of legitimate collective bargaining agreements that our union can exist and continue as a fundamental and integral part of the American democratic society."[8] Responsible adherence to contracts, not industrial democracy, was becoming the defining feature of labor unionism.

What is easy for us to lose sight of as present-day observers is the connection between these events and the broader class dynamics of the era. The ramifications were evident to left-wing journalists as well as in the behavior of the steel companies and the workers on the front line. The *Nation*, which was sympathetic to the liberal-labor alliance, understood this clearly.[9]

According to the editors, the steel magnates had engaged in a form of "class warfare," operating in a "thoroughly class conscious" manner, and had used "middle-class opinion in the form of vigilante and back-to-work movements" to break the strike. "Little Steel is freely using the language as well as the methods of civil war," the magazine editorialized. Moreover, the industry had "behind it the support of the nation's conservative press, which has united in an anti-union, anti-government campaign as self-righteous as it is misleading." Despite the *Nation's* stance, nothing short of assuring the mainstream media that the CIO was the labor equivalent of 4-H or the Rotary Club would win this segment of public opinion to labor's cause.

If media commentators missed the class character of public opinion, the La Follette Committee's investigators did not. In a curious addition to its examination of the Memorial Day incident, the investigators included a discussion of Gallup polls conducted by the American Institute of Public Opinion over the course of the previous six months that focused on the question of organized labor. In general, public opinion toward labor unions had remained strong, but public attitudes toward union tactics had become increasingly negative. While 76 percent of the respondents still supported labor unions, 67 percent believed that sit-down strikes should be prohibited, and 71 percent favored less unionism, rather than more. Equally telling, 57 percent agreed with the idea that the militia should be called in when a strike was threatened. What the Gallup polls point to is the corrosive effect of media reporting about the sit-down strikes and mass demonstration in the automotive and steel industries. Yet they also suggest the corresponding class dimension of the public's response to labor activism. "Although the C.I.O. has made important gains," the American Institute of Public Opinion noted, "its strike activities have alienated the sympathy of middle class opinion, *while keeping the support of the lower classes*" (emphasis added).[10] Similarly, the Institute's researchers noted a "sharp difference of opinion among economic groups on the labor question. People in the lower economic income brackets almost invariably take a position more favorable to labor unions than do the middle and upper classes." Additional polls, the Institute reported, "reveal that the middle and upper groups instinctively lean toward the conservative element in labor."

What do we today make of generalizations about public opinion turning against the labor movement in 1937? We know that Roosevelt was of the opinion that "the whole public" had lumped the labor unions and steelmakers together as troublemakers. What the president, as well as labor leaders Lewis and Murray, did not acknowledge was that public opinion was divided along class lines. Drawing on data from the American Institute of Public Opinion, the *Washington Post* reported that working-class support for the CIO had grown while middle-class opinion had turned against it.[11] While the report did not capture the ranks of the middle-class, Popular Front types who supported the movement, it still illustrated sharp class divisions about the legitimacy of the sit-down strikes. Coupled with union leaders' unwillingness to advocate tactics that might alienate the Roosevelt administration, the media assault damaged frontline morale, perhaps even more so than the brutality of the Chicago police force.[12] By tacitly accepting the official version of law and order in Chicago, the heads of the CIO conceded the point that organized labor had become a dangerous and subversive force.

In contrast, the United Packinghouse Workers focused intently on the problem of media representation. They sent a resolution to the governor of Illinois, condemning the "monopoly" that the mainstream media exercised at the newsstands, castigating it as a "direct violation of the peoples [sic] right of the freedom of press guaranteed under the Constitution." Their organizing committee demanded that the *Daily Times*, the *Daily Worker*, the *Catholic Worker*, the *People's Press*, and "all other publications sympathetic to labor" be made available at Chicago's newsstands.[13] The packinghouse workers understood better than Murray that the struggle was as much over ideas as it was about recognition.

This evidence should produce a note of skepticism about opinion polls or media reports that claim to represent "public opinion" regarding the labor movement. We should not be surprised that many middle- and ownership-class Americans disapproved of sit-down strikes. The wave of strikes to which the Little Steel Strike belonged was evidence of a working-class rebellion that most others did not understand and many would not support.

The Decline of a Popular Front Movement

It was anticommunism that ultimately acquitted the Chicago Police Department. It gained legitimacy when the federal government failed to squelch the hysteria of the Red Scare; its proponents would eventually turn this anticommunist impulse against the New Deal itself. The tactic was already evident during the Little Steel Strike, since the fervor of American anticommunism was not directed against foreign subversion, but organized labor. Since the New Deal had provided a mantle of legitimacy to the labor movement, those who sought to reverse its momentum invariably targeted its political patrons. In the 1940s, congressional committees investigating alleged communist subversion would also turn their attention to the film industry, academia, and the defense industries, but the labor movement would continue to be their primary target. Here, they would have the assistance of the newspapers, political pundits, and business associations that made their debut by attacking SWOC as a communist-riddled agent of Moscow. As historian Ellen Schrecker has written, "They inflated the significance of the communist presence in the labor movement, both to disguise their underlying hostility to unions and gain support for their opposition to the New Deal, which they believed sided with labor."[14] This network of anticommunist activists that was conspicuously evident in Chicago during the Little Steel Strike provided the underpinnings for the McCarthyism of the postwar period.

By the end of July 1937, the Chicago steelworkers' strike had effectively been lost. Plagued by inadequate finances, the indifference of the federal government, a phalanx of antiunion propaganda, police and National Guard violence, and collaboration between the Republic Steel Company and local officials, as well as the union's own failure to call for a general labor uprising, SWOC lost the Little Steel Strike. Even so, it carried on the fight for union organization in the plants and in the courts. Under the leadership of Joe Germano, SWOC pushed ahead in Chicago's Southeast Side and in Indiana Harbor, organizing workers on the shop floor despite vigilant company resistance and dwindling union funds. Respected SWOC leaders compelled foremen to join. That won them the loyalty of rank-and-file steelworkers. The renewal of the contract between SWOC and US Steel in the midst of the Roosevelt recession also bolstered the union's fortunes.

The Second World War created the conditions for a workers' breakthrough in Little Steel. Confronted by the demands of military production, the persistent activism of shop-floor organizers, and the relentless pressure of the federal government, the Little Steel companies finally capitulated. Republic Steel signed its first contract with the newly minted United Steel Workers of America (USWA) in 1942.[15] During the war, the USWA became even more closely tied to the Democratic Party and the federal government. As a reward for its commitment to maximizing wartime production, the Roosevelt administration established the National War Labor Board (NWLB), an agency designed to adjust union demands for higher wages and resolve grievances.

The NWLB granted several union demands that had been central to SWOC's organizing drive. These included the "maintenance of membership" clause, which ensured that workers who belonged to the union at the beginning of a contract remained union members throughout. The grievance apparatus also improved dramatically. As the NWLB eliminated the disparities between skilled and unskilled labor, the steelworkers gained measurable improvements from governmental safeguards and contract security. The board supervised wage increases, seniority mechanisms, and protections against unemployment that left industrial workers better off than before the war. These concessions stoked the fires of unionism, and membership positively skyrocketed. Moreover, the NWLB represented the achievement of a legalistic system of union representation and dispute resolution that realized the dreams of generations of labor-rights activists.[16]

Yet major labor unions such as the USWA accepted a narrower and more limited version of industrial democracy in return for wage and benefit improvements. In the process, the notion that workers should participate in governing the industries in which they worked became a distant memory. Instead of organizing new categories of workers and expanding protections for laborers in industries beyond the steel mills, the unions focused on their own members' limited interests. Rather than advocating for the interests of minorities and women, they grudgingly accepted court orders for equalization. In place of joining larger coalitions that challenged American unilateralism abroad, they supported American foreign policy in the Cold War, in Vietnam, and in Latin America.

There is no doubt that the steelworkers' union achieved concrete improvements for their members. Through collective bargaining agreements, they raised wages for those on the shop floor; created seniority systems; and won generous benefit packages, safeguards against unemployment caused by automation, and a system of industrial jurisprudence that mitigated the arbitrary authority of the steel industry's management. These achievements also raised the wage levels in their local labor market and provided a mechanism for redistributing corporate profits in the form of wages, thus contributing to the economic health of their communities. It's also important to note that the CIO did not relinquish its commitment to more fundamental political and social reform simply because of what happened in the Memorial Day

Massacre. It would take the demands of wartime production, the reaction of the business community, and anticommunism to accomplish that task.

What the episode in Chicago did do, however, was to erode the idea that the maldistribution of wealth and power was a public issue that required political solutions. It shifted the union movement toward greater centralization and far more caution. It did away with the belief that mass protests based on broad alliances could challenge economic injustices. It reinforced the idea that companies and unions should determine wages and pensions in closed-door sessions, shut off from the interests of a broader society. By the 1950s, the CIO had lost the sense that it belonged to a larger united front that still had to address fundamental issues of political and economic inequality.[17]

The United Steel Workers of America would continue to collaborate with political elites throughout the course of the twentieth century. It's important to remember that Chicago's Mayor Kelly ultimately capitulated to the steelworkers. Fearing a backlash from industrial workers, it was Kelly who ensured a more hospitable environment for union organizing in that city. The political forces that shaped the New Deal also changed Chicago's local political landscape. The relationship between organized labor, the Chicago machine, and the Democratic Party paid enormous dividends. Steadily increasing wages geared to the rising cost of living, substantially improved benefits, job safeguards, grievance-resolution mechanisms, and protections from technological unemployment substantially improved the lives of millions of steelworkers. Concerted labor organizing, in conjunction with the New Deal's accomplishments that provided a measure of security for average Americans, produced the famed "middle class" of the postwar era. (Here we should be careful, however, since driving a tail-finned car and owning a washing machine did not mean that an employee had gained control over production. Earning a decent living did not make one a member of the middle class, but this was certainly the terminology used at the time to describe industrial laborers who had "made it.") CIO leaders and the authority of the federal government finally unionized Little Steel.

But when political fortunes changed, organized labor suffered the consequences. Tied too closely to the Democratic Party, labor lacked the political independence to challenge the party's policies. In issues ranging from escalation of the war in Vietnam, to environmental despoliation, to an overemphasis on consumption rather than national economic planning, organized labor functioned as the junior partner in a program designed to expand American consumerism and project American influence abroad. In the late 1940s, as the nation embraced the Cold War, the CIO would cut itself loose from its more progressive and left-leaning supporters in order to protect itself from McCarthyism. That choice would only accelerate the union's drift toward a more conservative labor movement. The mainstream unions would have many victories: some in the field of civil rights, and most in the area of raising the living standards of its members. They would also have to fight for many of those gains. In 1946 and 1959, steelworkers would engage in massive walkouts that clearly indicated the steel industry had reluctantly and temporarily accepted a labor-management accord. Still, these achievements would benefit not only union members, but the larger communities in which they lived and worked.

Yet most of the mainstream labor unions would not recover their vision of transformative social justice until the 1990s. Devastated by deindustrialization and corporate-led antiunion campaigns, they fought defensive actions just to protect existing contracts and a dwindling membership. In the case of the steel industry, foreign competition dealt it a terrific blow. Federal trade policies that permitted foreign imports but refused to demand reciprocal opportunities for American manufacturers exacerbated the problems plaguing the industry. These included the failure on the part of the steel companies' management to promote innovation, adjust to new market demands, and improve productivity. Yet the steelworkers themselves must bear some of the responsibility for the industry's misfortunes. Steadily increasing spirals in workers' wages and benefits in a period of escalating competition intensified the crisis facing the American steel industry.

Political maneuvering in the postwar era only exacerbated the problem. By opposing a national health-care program and failing to pursue the policy of full employment that enjoyed widespread support during World War II, political elites and union leaders created a privatized system of health care, welfare, and collective bargaining that was highly unstable. It was sustainable only as long as American corporations dominated the international economy. Once that privileged position ended, companies had to slash benefit and wage packages in order to restore profitability. Confronting a growing conservative movement that rationalized this corporate counterattack in the name of individual freedom and disparaged social safety-net programs as a drag on economic efficiency, workers found themselves cut adrift.

Rethinking the Little Steel Strike

These more recent developments bring us back to the lost opportunities of the New Deal and the social democratic impulse underlying the labor rebellion of 1937. Historians of the Little Steel Strike have tended to perpetuate middle-class opinion from that era, which held that the mass demonstrations were misplaced. Considering the effectiveness of the direct-action tactics and the benefits of a broader alliance than in previous strikes, there is good reason to question this assertion. SWOC leaders may very well have failed to cultivate sufficient support for the union effort in the Little Steel mills. That was certainly the conclusion that journalist Benjamin Stolberg arrived at when he surveyed the collapsing strike at the end of July 1937. His claim that there was "no tradition of militant labor" in the mill towns involved in the strike was exaggerated, but his observation that the steel industry held a dependent and vulnerable workforce in its grip rings true.[18] "In such isolated and terrorized communities," Stolberg noted, "an inexperienced, backward, and confused working class can be organized only through a thorough and patient process, by organizers who know the industry, the psychology of the workers, the whole social lay of the land." It was for this reason that Stolberg regretted the "easy victory" of SWOC in organizing US Steel, because it encouraged the notion that a similar victory in the Little Steel mills was inevitable.

Equally regrettable is the limited scope of those gains, because it is in such moments of public rebellion that citizens engage in a more profound examination of social inequality. Not every demonstration produces the kind of intellectual and moral ferment that resulted from the events in Chicago in 1937. Some protests are motivated by narrow self-interest, unreflective anger, and racial intolerance. We need look no further than the "hate strikes" launched by autoworkers in Detroit during the Second World War to remind ourselves that not every act of resistance is virtuous. In the atmosphere of the 1930s, however, labor protests often provided an occasion for a more thoroughgoing critique of American society.

This is what the failure of the Little Steel Strike produced: a more profound, albeit brief, exploration of possibilities beyond "business as usual." What was percolating up from the streets, the classrooms, the union halls, the churches, and the community centers of Chicago was a set of beliefs that ran counter to the assertion that private enterprise should govern American society. Yet those convictions also challenged the sort of unilateral authority exercised by steel executives and union leaders alike. A profoundly democratic impulse was evident at this moment, and it linked the activists of the 1930s with those in the Gilded Age who advocated for a more cooperative commonwealth. In some sense, then, the leaders in the steelworkers' union were working at cross-purposes to this larger social movement.

These leaders also undermined the possibility of winning the Little Steel Strike. Mass demonstrations may not have compensated for insufficient union organization of laborers inside the steel mills. That certainly seemed to be Benjamin Stolberg's opinion. Yet by capitulating to the mainstream media's censure and the political expediency of the Roosevelt administration, SWOC leaders practically ensured that the walkout would collapse. Instead of mobilizing a wide-scale movement in defense of the egalitarian ideals of the CIO, the union's upper echelon chose to retreat into "respectability." Rather than confronting the Roosevelt administration directly, as John L. Lewis did, the steelworkers' union capitulated to the argument that the existing Democratic Party was the best they could hope for. The promise of a contract won out over the idea that this was a moment in which to crystallize the labor movement. Although the guns of the Chicago Police Department, the relentless hostility of Tom Girdler, and the betrayal of the Roosevelt administration played the largest parts, the CIO itself impeded the momentum of the 1937 sit-down movement toward a more fundamental realignment of economic power in America. As John L. Lewis observed: "The men in the steel industry who sacrificed their all were not merely aiding their fellows at home but were adding strength to the cause of their comrades in all industry. Labor was marching toward the goal of industrial democracy and contributing constructively toward a more rational arrangement of our domestic economy." The 1937 strike did more than delay collective bargaining in the steel industry; it undermined the drive for a "more rational arrangement" of American society.[19]

Yet why didn't rank-and-file workers do more to determine the course of the strike? Workers in the auto, rubber, and textile industries had no trouble defying their unions' leaders. Answering this query may require us to return to Benjamin Stolberg's observations about "the inexperienced" and "confused" working class that populated these "isolated" steel-mill villages. It raises questions about the political culture within steelwork, which many historians have noted was more conservative than in the rubber, meatpacking, auto, and electrical industries. It brings up issues about the legacy of the 1919 strike: the persistence of ethnic and racial divisions. But it also leads to a larger concern: had workers become too devoted to Roosevelt? Too willing to submit to his authority out of love, fealty, and fear of the alternatives? What we do know is that the US Steel victory occurred without a strike, coming through negotiations that left rank-and-file workers on the sidelines. The result predisposed workers to expect a breakthrough in Little Steel that did not require a monumental, protracted mobilization. Since the beginning of their organizing drive, steelworkers had relied extensively on the efficient leadership of SWOC, which had exercised as much top-down hierarchical control as it could. When the moment came for rank-and-file workers to challenge their leaders' decisions and control, their prior experience and conditioning offered them little guidance. What is most remarkable about the Little Steel Strike, then, considering the political and historical context of work in the steel industry, *is that a popular movement of social protest developed at all.*

But what about the middle-class activists? Why didn't they do more to sustain the movement for social reform? They may not have been able to challenge an extremely popular president, but they might have expended more of their intellectual and financial resources on winning the steel-industry strike, encouraging mass-action tactics, and defending a vision of social democratic reform. Does the Memorial Day episode illustrate the divisions between middle- and working-class Americans over strategies of social reform, the objectives of social protest, and sheer self-interest? Does it highlight the limits the middle class imposed on social activism? The political atmosphere was certainly changing in 1937. Roosevelt's court-packing scheme and a growing antipathy toward the sit-down strikes contributed to this class's reaction against the New Deal. Yet we have good reason to wonder whether the unwillingness of middle-class progressives to do more in defense of the steelworkers' unionization drive compounded the drift away from a more expansive New Deal.

In the study of the past, however, what *did* happen must always take priority over speculations about what *might* have happened. What did happen was that the pace of New Deal reform slackened considerably. And although the CIO had not abandoned its dreams of playing a larger role in the economic and political life of the nation in 1937, after World War II it moved further toward dampening rank-and-file enthusiasm, at the same time that it accepted the subordinate position assigned it by corporations and the cold war state.

Historical analysis is also a question of perspective. By shifting our focus from SWOC headquarters and the boardrooms of Republic Steel to the larger protest that the Little Steel Strike unleashed, a different view of the labor movement in the 1930s appears. We begin to see how nonstriking trade unionists, social workers, housewives, high school and college students, activist clergy, socially conscious academics, liberal professionals, and sympathetic writers participated in this labor movement, which clearly extended beyond the boundaries of the union. Long before the steelworkers carried their first picket signs in front of a Republic Steel plant, unemployed black and white workers had mobilized in a militant challenge to the social order.[20] Students and left-wing intellectuals joined that movement in strength. From the Bonus Marchers of 1932 to the housewives who organized their neighborhoods and launched the meat-industry strikes in 1935, working-class Americans shaped the social and political vision that would become the New Deal.

By expanding our field of vision beyond union leaders, Washington politicians, and corporate executives, we gain perspective on why a dispute between a few steel companies and their employees mattered to anyone else. Despite the persistence of racial intolerance and sexism, African Americans, Latinos, and women from across the ethnic spectrum made progress. Through their activism on the streets of Chicago's Southeast Side and in the women's auxiliaries, they took strides toward greater social equality, political inclusion, and personal dignity. They played vital roles in the entire Little Steel Strike, not just on Memorial Day 1937. From a larger perspective, we can see how this labor upheaval contested the racial and gender norms of twentieth-century America. From a longer time frame, the continuities between the Unemployed Council movement and the march to Republic Steel's Burley Avenue plant become visible. The Little Steel Strike precipitated a community uprising that raised fundamental questions about the morality and decency of America's economic order.[21]

Many of the people who participated in that march believed that the "labor question" was the fountainhead for resolving the social and political inequalities that had plagued industrializing America. They sincerely believed that the drive for unionization was at the center of the effort to democratize

American society. In this decisive moment during the Great Depression, a movement linking industrial laborers and the middle-class left challenged the steel industry, a powerful symbol of corporate dominance in America. When we look beyond the tactical maneuvering of SWOC or the political calculations of Republic Steel, then we begin to perceive this dynamic. What we are observing are the forces behind class conflict in American history.

Epilogue
Rethinking the Massacre

LOOKING BEYOND THE CARNAGE of Memorial Day 1937 in Chicago, we begin to see this point in time as something more than a matter of victims and brutality and bloodshed. It was one of the moments in the 1930s when a progressive alliance began to live according to a set of ideals that rejected the primacy of profits and private property over human rights. Those cooperative and egalitarian ideals represented a profound break from the individualistic ethos of the 1920s. We should be careful not to romanticize this brief period, however: the workers *did* lose the strike; they did *not* form—or even aspire toward—a utopian commune; and they did *not* lose sight of their material interests. At the same time, we should not assume that, since the steelworkers ended up in a conservative, often racially intolerant, business-friendly labor union, this was the mindset of those workers in 1937.[1] The same could be said for a large percentage of the other laborers who joined the steel union's struggle in Chicago's Southeast Side. Just as importantly, liberal activists saw workers as allies in an important cause, not opponents whose decision to go on strike created an inconvenience for them.

We need to acknowledge that the Memorial Day demonstration was a significant example of the kind of mass protest that was necessary and often effective in the 1930s. In trying to exonerate the marchers of any wrongdoing, historians have tended to minimize the groundswell of popular indignation that fuelled the Little Steel Strike. Mass demonstrations were neither passive nor merely symbolic affairs. Even when they were nonviolent, which was certainly the initial intent of the demonstration at Republic Steel's Burley Avenue plant, mass protests became exercises in civil disobedience that challenged the claim that the police defended law and order. The size of the Memorial Day march clearly alarmed the police. Some of the participants probably did plan to enter the plant, some carried crude weapons that day, and some hurled objects as well as invectives at the bluecoats. Considering the previous week's harassment of strikers by the police and the experience of social protest in the early 1930s, none of this should be surprising.

The Memorial Day demonstration was planned neither as a violent nor an aggressive event, but insisting on the innocence of the marchers obscures the place of this protest in the working-class movement of the 1930s. Participants believed that an enormous amount was at stake. If the labor struggle of the 1930s demonstrated anything, it was the willingness of workers to take enormous risks to achieve what were often very limited demands. In the midst of those struggles, housewives and other women, nonstriking supporters, and workers fighting for their livelihoods responded in kind to police provocations or to attacks by company thugs. By focusing on police brutality and ascribing a lack of culpability to the marchers, historians have obscured the percolating discontent that exploded into mass protest at Chicago's Republic Steel plant. This was not token activism, but a powerful expression of the contest between average Americans and those who were determined to preserve existing hierarchies. The intransigence of the Little Steel executives in the face of the National Labor Relations Act demonstrated that legislation would not be enough. If the federal government would not intervene to protect workers, they would once again have to turn to collective action.

What complicates the issue even further is the fact that the history of labor in the 1930s is studded with incidents where civil disobedience *worked*. As labor analyst Joe Burns has pointed out, workers in the 1930s understood that to prevent companies from treating them like commodities, they would have to halt the production line.[2] If they did not, the firms would simply replace them with other semiskilled laborers who would work for even less. That would further erode wages for *all* workers and tighten the unilateral control of big business over the economic system on which average Americans now depended. In order for laborers to stop production, they either had to prevent nonstriking workers from getting into the plant, or compel those nonstriking workers inside the plant to leave. Those rudimentary calculations emerged in every strike situation in which a company had thwarted the Wagner Act. That was certainly the case in strike areas in Ohio, where labor journalist Mary Heaton Vorse had seen class warfare in the streets. It was also evident, in Chicago's Southeast Side, where striking steelworkers fought police officers in the week preceding the Memorial Day incident.

The objective of all of this was to equalize wages across entire industries. It was only by standardizing wages that labor unions could remove this factor from the cut-and-thrust between labor and management and thus impede the ruthless pitting of workers against one another.[3] If companies wanted to maximize profits, they would have to do so by manufacturing better products more efficiently, not by driving wages down to subsistence levels. The movement to organize labor was intended not only to prevent American workers from being subjected to abuse and peril inside the factories, but also to ensure that businesses could not engage in the relentless pursuit of profit at the expense of the livelihoods of those who did the work. That meant marching, demonstrating, engaging in sit-ins, and fighting back. It also meant relying on the support of others who saw the laborers' fight as their own.

Despite a tradition of community resistance, there is equally compelling evidence that elites have used instances of violent retaliation against them and their practices to discredit social reform. Workers' opponents took any occasion where there was physical resistance to build the case that organized labor was *inherently* violent, disorderly, and a threat to social stability. The history of American social protest is littered with incidents in which government and business collaborated not only to inflict greater violence on dissenters, thus winning the physical struggle, but also to twist even feeble demonstrations of resistance into acts of unbridled lawlessness, thus eroding the protestors' claims to legitimacy. Mainstream media has frequently played a supporting role in the effort to discredit social protest. When alternative news sources, such as the CIO News and the Federated Press, contested the dominant narrative of "respectable" business versus "fanatical" labor, their challenges proved to be less effective. It is highly unlikely that Chicago's leading newspapers would have reported on the Little Steel Strike as they did had the demonstrators been composed of choirboys from Holy Name Cathedral on downtown Chicago's North Wabash Avenue.

Violence, however defensive and spontaneous, has consistently provided the pretext for state suppression of dissent in American history. Working-class activists absorbed this tradition. One of the key reasons why they adopted the sit-down tactic was that it was extremely effective in neutralizing the kind of violence that had so often been turned against workers. Of course, that is also why big business and its allies in Congress fought so hard to prohibit this form of labor activism. It is important to recall that, in the early moments of the Little Steel Strike, workers at Republic Steel's Burley Avenue plant did try to organize a sit-down strike by occupying the mill. The emerging civil rights movement would later demonstrate the strategic benefits of a principled commitment to nonviolent civil disobedience. While it's important to understand the limits of militant tactics, then, it's also crucial to comprehend why workers adopted them in the first place. In 1937, Edward Levinson wrote that the wave of sit-down strikes that year was "labor's reaction to the frustration and hopes it placed in the New Deal."[4] Speaking specifically of the Wagner Act, Levinson observed that "union men cannot be expected to see their leaders picked off, their wages held down, the machines speeded up to inhuman pace, and meanwhile await patiently the outcome of court processes by which a law already enacted and signed is being balked on advice of conservative lawyers to their ready clients." This perspective was lost in the public opinion polls, which framed the questions for those who answered the surveys. Those same polls also provided minimal insights into how well respondents understood the issues for which their opinions had been solicited. The topics in the polls did little to account for the impact of negative media reporting on the labor movement in the months preceding a strike. This was the media environment that confronted SWOC as it tried to keep the Little Steel Strike alive in July 1937.

When a culture of labor activism thrived in the United States, it at least counterbalanced the social prestige of business leaders and checked the prerogatives of capitalism. It mitigated the tendency to privilege private property over the human rights of those whose muscles built America. As the opponents of economic and racial justice asserted their dominance in the postwar years, the idea that workers were in the forefront of a movement for social democracy deteriorated. At the same time, the cultural authority of corporate America surged. In the polarizing atmosphere of the post-1947 cold-war era, alternative media outlets sympathetic to progressive labor unionism diminished steadily. It became easier and easier for mainstream media to make the case that labor militancy equaled "labor violence," that strikes threatened the public interest, and that a small group of labor "agitators" were responsible for workplace activism. In the neoliberal atmosphere that has dominated the United States since the 1980s, the courts, the federal government, most state governments, and a range of firms specializing in "union avoidance" have joined the media in stigmatizing labor unionism. For all intents and purposes, the labor movement that George Patterson and James Stewart and Lucille Koch built is but a memory. Yet memory is a powerful thing.

What this discussion should do is raise some of the questions that have occupied advocates of social justice since the Gilded Age. Is collective protest enough to effect social change? Does the mainstream political system offer an avenue for the realization of working-class aspirations? When is going on strike necessary, morally legitimate, and politically viable? What are the philosophical foundations of labor unionism? Does a workers' movement have a place in our globalized, competitive economy? Do unions still represent the voices of the dispossessed and disadvantaged?

Opponents of labor unionism acknowledge the virtue of human rights, but they would argue that it's the rights of workers who *do not* wish to join that are circumvented when unions win. According to this view, union contracts unfairly bind the hands of business owners who must operate in a competitive economy where profit margins are often slim. Instead of promoting economic growth, they supposedly stifle job creation by sending business owners searching for ways to lower their operating costs. Where union contracts do prevail, the cost of what they provide for workers is simply passed on to consumers, many of whom do not earn union wages. Furthermore, exponents of this view argue that the collectivism inherent in the union movement goes against the grain of an American tradition that cherishes individual freedom and self-improvement. Critics of labor unionism assert that workers should not be forced to join a union against their will. (Defenders of labor unions point out that *nowhere* are workers compelled to join a union. In right-to-work states—which don't guarantee employment for those looking for work, but rather have laws stating that membership in a union is not required as a hiring condition or as a way of keeping a job—employees who do not sign a union card enjoy the same wages and benefits gained through collective bargaining, yet have no obligation to contribute dues to cover all of the expenses that go into negotiating a contract.) In essence, this perspective emphasizes the idea that unions function as job-control monopolies that decreases economic efficiency. Recent critics of the labor movement have also pointed out that, despite the rhetoric of equality and dignity, unions have a long and sordid history of discriminating against minorities and women.

Perhaps the most effective argument offered by present-day advocates of the "right to work" is that, in an era of corporate globalization, the United States can no longer afford labor unions. In a global "race to the bottom" (in which nations, and even states within nations, compete against each other for private investment by consistently lowering labor costs and social protections), the preservation of union prerogatives only makes American labor more expensive and less appealing to the fleet-footed corporations participating in the competition. Those espousing these arguments point to evidence showing that auto manufacturers and other industries have deliberately sought out right-to-work states, thus creating jobs and opportunities for people who simply want to work. They contend that it is not right to permit only certain cadres of workers to have the benefits of a union contract when states are forced to cut social services and most workers face growing job insecurity. In an era of diminished expectations, many critics, including those from the working class, view labor unionists as a privileged elite who—according to prevailing thought—have grown lazy, unproductive, and aloof from the problems confronting nonunion workers.

Proponents of labor unionism would argue that, in a period of deteriorating wages, rising unemployment, and grotesque disparities in the distribution of wealth, the United States cannot afford *not* to protect the rights of labor. Additionally, they would submit that labor unions set wage and workplace standards at the levels to which they need to rise for *all* workers. Without decent-paying jobs, and without the forced redistribution of corporate profits in the form of wages guaranteed by union contracts, communities would be much worse off than they already are. Labor supporters in the 1930s made the same case for the economic benefits of unionization. By ensuring that companies allocated their profits more equitably, labor unions countered the maldistribution of wealth that many observers believed was a leading cause of the Great Depression. Economists who have supported the New Deal contend that by increasing workers' buying power, labor unions help stimulate economic growth and reduce the concentration of wealth in just a few hands.

These are questions for contemporary students, business leaders, workers, and citizens to debate. But they can only responsibly discuss such issues by examining America's past. Stripped of myth, misconception, and political bias, that study will reveal a lineage of labor activists who have been in the forefront of progressive social change since the nineteenth century. From racial equality to international peace to environmental consciousness and women's rights, labor unions have been inextricably connected to the movement for social justice. Even in the periods when mainstream unions seemed to have entered the deep freeze of conservative thought and practice, individual unions have championed racial equality, the rights of migrant workers, and the benefits of economic planning for those affected by job losses due to restructuring. Historically, labor unions and collective movements of protest have provided the key instruments to challenge the dominance of the few over the many.

That is precisely why those committed to maintaining the status quo oppose them so virulently. "Unions not only challenge private power," Peter Zwiebach has written, but "they offer a possibility of a change in value systems away from an individualistic ethos to one more centered on solidarity and justice. Unions offer the potential to curb economic power in significant ways, and their mere existence can limit private economic power in ways that no other human rights organization can."[5] Like the workers and community members who joined the movement for industrial democracy in the steel industry, more and more of today's advocates for organized labor see it as a means of restoring basic human rights. The most fundamental is the right to earn a decent living. Labor supporters have argued that a worker's right to a fair wages, to dignified work, and to decent working conditions should carry the same moral weight as the human rights that we claim to revere. In 1948, the United Nations issued a Universal Declaration on Human Rights that made the same case. In 1992, the United States signed the International Covenant on Civil and Political Rights, which recapitulated the right to join a labor union as an expression of the underlying right to freedom of association.[6]

Advocates have also pointed to the Bill of Rights as the foundation for the claim that collective bargaining should be protected. From this perspective, the Constitution protects the rights of freedom of assembly and freedom of speech, both of which are subverted when the efforts of workers to organize are obstructed. Furthermore, paying people inadequate wages, exposing them to dangerous working conditions, abandoning them to arbitrary managers, and rendering them permanently insecure in an extremely prosperous nation violates every element of the American creed. Those arguing this point of view have made the case, convincingly, that most workers—both now and in the 1930s—resisted unionization out of fear of losing their jobs, not out of principle. Proponents of labor unionism—in no matter what era—have argued that workers possess a democratic right to self-determination in the workplace. They have the right not to be treated like commodities to be bought, sold, and tossed away when no longer needed.

In light of the questions posed earlier in this chapter, understanding how Americans answered them at a given moment in the nation's history—in this case, 1937—can help us appreciate that such issues are not simply abstract questions for academic debate. Equally importantly, the story of the Memorial Day events on Chicago's Southeast Side in 1937 reminds us that these problems have a history. The past should inform the debate over labor unionism, since the struggle for workplace democracy has defined pivotal moments in the American experience. The early twenty-first century is not the 1930s, so we must be careful about understanding people and their behaviors within the ethos of their time, and not impose our expectations, views, and needs on them. That's the constant challenge of writing history. Nonetheless, the arguments marshaled in favor of and against union representation sound remarkably similar in 1937 and today. Stripped of their historical peculiarities, some of the questions generated by the crisis of the Great Depression do indeed transcend that era. That should come as no surprise, since a fundamental aspect of that period is the perennial question, what does a society supposedly committed to the principle of democratic equality really look like? If some of the above questions resonate beyond the 1930s, perhaps some of the answers do as well.

NOTES

PROLOGUE: The Making of a Memorial

1. Robert Bussel, "'A Love of Unionism and Democracy': Rose Pesotta, Powers Hapgood, and the Industrial Union Movement, 1933–1949," *Labor History* 38 (Spring/Summer 1997): 210.
2. Roger Biles, "Edward J. Kelly: New Deal Machine Builder," in Paul M. Green and Melvin G. Holli (eds.), *The Mayors: The Chicago Political Tradition*, 3rd ed. (Carbondale: Southern Illinois University Press, 2005), 111–15.
3. Mary Heaton Vorse, *Labor's New Millions* (New York: Modern Age Books, 1938), 288.
4. Ibid., 126.

CHAPTER ONE: From Crisis to Confrontation

1. "The U.S. Steel Corporation: III," *Fortune* 13 (May 1936): 134.
2. *Violations of Free Speech and Rights of Labor: Report of the Committee on Education and Labor, Private Police Systems . . . Harland County, Ky . . . Republic Steel Corporation* (New York: Arno Press and New York Times, 1971 reprint, originally published 1939), 4–14.
3. See Rosemary Feurer, *Radical Unionism in the Midwest, 1900–1950* (Urbana: University of Illinois Press, 2006), 31–34; Randi Storch, *Red Chicago: American Communism at its Grass-Roots, 1928–1935* (Urbana: University of Illinois Press, 2007), 99–129; Frances Fox Piven and Richard A. Cloward, *Poor People's Movements: Why They Succeed, How They Fail* (New York: Pantheon Books, 1977), 53–55, 58–59; Michael Goldfield, "Worker Insurgency, Radical Organization, and New Deal Labor Legislation," *American Political Science Review* 83 (Dec. 1989): 1257–82.
4. Joseph A. McCartin, "Reframing US Labor's Crisis: Reconsidering Structure, Strategy, and Vision," *Labour/Le Travail* 59 (Spring 2007): 146–47; Nelson Lichtenstein and Howell John Harris, "Introduction: A Century of Industrial Democracy in America," in Lichtenstein and Harris (eds.), *Industrial Democracy in America: The Ambiguous Promise* (New York: Cambridge University Press, 1993), 5–6; Lichtenstein, *State of the Union: A Century of American Labor* (Princeton, NJ: Princeton University Press, 2002), 30–32.

5. Michael Denning, *The Cultural Front: The Laboring of American Culture in the Twentieth Century* (New York: Verso Press, 1996), 4–10.

6. Ian McKay, *Rebels, Reds, Radicals: Rethinking Canada's Left History* (Toronto: Between the Lines, 2005), 140–41.

7. David Kennedy, *Freedom From Fear: The American People in Depression and War, 1929–1945* (New York: Oxford University Press, 1999), 86–88.

8. Lizabeth Cohen, *Making a New Deal: Industrial Workers in Chicago, 1919–1939* (New York: Cambridge University Press, 1990), 238–49.

9. Ibid., 261–62; Rick Halpern, *Down on the Killing Floor: Black and White Workers in Chicago's Packinghouses, 1904–1954* (Urbana: University of Illinois Press, 1997), 108; Storch, *Red Chicago*, 111–15.

10. Ronald Edsforth, *The New Deal: America's Response to the Great Depression* (Malden, MA: Blackwell, 2000), 105.

11. Steve Nelson, James Barrett, and Rob Ruck, *Steve Nelson, American Radical* (Pittsburgh: University of Pittsburgh Press, 1981), 76.

12. Storch, "Shades of Red: The Communist Party and Chicago's Workers, 1928–1939," PhD diss., University of Illinois at Urbana-Champaign (1998), 85–88.

13. Storch, *Red Chicago*, 111–13; Storch, "Shades of Red," 84–89.

14. Joshua Bartlett Lambert, *"If The Workers Took a Notion": The Right to Strike and American Political Development* (Ithaca, NY: ILR Press, 2005), 133, 203.

15. Ibid., 98–100.

16. James Gray Pope, Peter Kellman, and Ed Bruno, "The Employee Free Choice Act and a Long-Term Strategy for Winning Workers' Rights," *Working USA: The Journal of Labor and Society* 11 (Mar. 2008): 133–35; Pope, "How American Workers Lost the Right to Strike, and Other Tales," *Michigan Law Review* 103 (2005): 527–34.

17. Robert Zieger, *The CIO, 1935–1955* (Chapel Hill: University of North Carolina Press, 1955), 50.

18. George Patterson autobiography, bk. 1, p. 138, box 9, folder titled "Autobiography," George Patterson Papers, Chicago Historical Museum [hereafter abbreviated as CHM].

19. Feurer, *Radical Unionism in the Midwest*, 236.

20. Irving Bernstein, *The Turbulent Years* (New York: Houghton Mifflin, 1970), 467–71.

21. "Steel Victory—and After," *Nation* 144 (Mar. 6, 1937): 286.

22. Art Preis, *Labor's Giant Step: Twenty Years of the CIO* (New York: Pioneer Publishers, 1964), 65.

23. Walter Galenson, *The CIO Challenge to the AFL: A History of the American Labor Movement, 1935–1941* (Cambridge, MA: Harvard University Press, 1960), 93; Bernstein, *Turbulent Years*, 468; David Brody, *Workers in Industrial America: Essays on the 20th Century Struggle* (New York: Oxford University Press, 1982 reprint, originally published 1980), 103–6.

24. Zieger, *CIO, 1935–1955*, 58–59; Lichtenstein, *State of the Union*, 52; Galenson, *CIO Challenge to the AFL*, 95; Halpern, *Down on the Killing Floor*, 126–27.

25. Letter of Chicago-area SWOC lodges to the "Honorable President of the United States, Members of the Senate of the United States, Members of the House of Representatives of the United States, Members of the Supreme Court, Members of the United States Labor Relations Board," Jan. 17, 1937, in *The CIO Files of John L. Lewis, Part 1: Correspondence with CIO Unions, 1929–1962* (Frederick, MD: University Publications of America, 1988), microfilm [hereafter cited as John L. Lewis Papers (microfilm)].

26. Vorse, *Labor's New Millions*, 113; Edward Levinson, *Labor on the March* (New York: University Books, 1956), 200.

27. Vorse, *Labor's New Millions*, 111.

28. Tom M. Girdler, in collaboration with Boyden Sparkes, *Boot Straps: The Autobiography of Tom M. Girdler* (New York: C. Scribner's Sons, 1943), 226.

29. David Brody, "The Origins of Modern Steel Unionism: The SWOC Era," in Paul F. Clark, Peter Gottlieb, and David Kennedy (eds.), *Forging a Union of Steel: Philip Murray, SWOC, and the United Steelworkers* (Ithaca, NY: ILR Press, 1987), 24.

30. Bernstein, *Turbulent Years*, 477–78.

31. Donald S. McPherson, "The 'Little Steel' Strike of 1937 in Johnstown, Pennsylvania," *Pennsylvania History* 39 (1972): 220.

32. *Violations of Free Speech and Rights of Labor: Report of the Committee of Education and Labor, Report No. 151, Labor Policies of Employers' Associations, Part 4; The "Little Steel" Strike and Citizens' Committees* (Washington, DC: US Government Printing Office, 1941), 118.

33. Brody, *Workers in Industrial America*, 109.

34. Ibid., 120.

35. National Labor Relations Board, "In the Matter of Republic Steel Corporation and Steel Workers Organizing Committee, Case No. C-184," Oct. 18, 1938, 84.

36. *Labor Policies of Employers' Associations*, 120.

37. Dorothy Patterson form letter to "Dear Friend," Nov. 23, 1936, box 9, folder titled "Women's Auxiliary," Patterson Papers. Also see "Auxiliaries Active in Enlisting Support for Organizing Drive," *Steel Labor*, Nov. 20, 1936.

38. Dorothy Patterson to "Dear Friend," Nov. 23, 1936, Patterson Papers.

39. Elizabeth Faue, *Community of Suffering and Struggle: Women, Men, and the Labor Movement in Minneapolis, 1915–1945* (Chapel Hill: University of North Carolina Press), 12–13, 122–25, 190–92.

40. "The Men's Corner," *Women in Steel* 1 (Jan. 1937): 1, Samuel Evett Papers, Calumet Regional Archives, Indiana University Northwest.

41. Annelise Orleck, "'We Are That Mythical Thing Called the Public': Militant Housewives during the Great Depression," *Feminist Studies* 19 (Spring 1993): 149–50, 167–68; Mary Eleanor Triece, *On the Picket Line: Strategies of Working-Class Women during the Depression* (Urbana: University of Illinois Press, 2007), 39–46.

42. John L. Lewis, "Industrial Democracy," NBC radio broadcast, Dec. 31, 1936, CIO 41, Publication No. 9, Jan. 1937, in Harold J. Ruttenberg Papers, box 4, folder

9, "CIO, 1937–1958," Historical Collections and Labor Archives, Special Collections, Paterno Library, Pennsylvania State University [hereafter abbreviated as HCLA].

43. James R. Barrett, "Rethinking the Popular Front," *Rethinking Marxism* 21 (Oct. 2009): 531–50; Michael Denning, "Afterword: Reconsidering the Significance of the Popular Front," *Rethinking Marxism* 21 (Oct. 2009): 551–55.

CHAPTER TWO: The Rising Tide of Rebellion

1. James Kollros, "Creating a Steel Workers Union in the Calumet Region, 1933 to 1945," PhD diss., University of Illinois at Chicago (1998), 213–15.

2. Minutes of SWOC fieldworkers' meetings, Apr. 26, May 13, 20, and 24, 1937, District 31, United Steel Workers of America, box 124, folders 124–26, CHM. The mood of militancy as well as the enthusiastic support of the Chicago-area steelworkers for CIO and SWOC leaders is equally evident in a resolution that the Calumet District lodges of the Amalgamated Association of Iron and Steel Workers passed in April 1937. James Stewart of Local 65 signed on behalf of area lodges, resolving that they would "give the Steel Workers Organizing Committee our loyal support and cooperation in the fulfillment of our obligation to the enforcement of our joint contract" and the organization of the remaining steelworkers. See "Resolution," Apr. 14, 1937, John L. Lewis Papers, Part 1 (microfilm).

3. Louis Leotta, "Girdler's Republic: A Study in Industrial Warfare," *Cithara* 11 (1971): 50.

4. "Bulletin No. 3, Chicago Department of Law Opinion Bulletin, Picketing," *Violations of Free Speech and Rights of Labor: Hearings Before a Subcommittee of the Committee on Education and Labor, United States Senate . . . Part 15-D, The Chicago Memorial Day Incident: Industrial Munitioning*, November 18, 1937, 75th Cong., 2nd Sess. (Washington, DC: US Government Printing Office, 1938), 6752 [hereafter cited as *LSC Part 15-D*].

5. Carl Swidorski, "The Courts, the Labor Movement, and the Struggle for Freedom of Expression and Association, 1919–1940," *Labor History* 45 (Feb. 2004): 77.

6. "Exhibit 3421-A: Affidavit—Paul Glaser," *LSC Part 15-D*, 6751.

7. Ibid.

8. Richard Linberg, *To Serve and Collect: Chicago Politics and Police Corruption from the Lager Beer Riot to the Summerdale Scandal, 1855–1960*, 2nd ed. (New York: Praeger, 2008), xiv.

9. "The Kelly-Nash Political Machine," *Fortune* 14 (Aug. 1936): 120.

10. Bernstein, *Turbulent Years*, 483; Donald Sofchalk, "The Little Steel Strike of 1937," PhD diss., Ohio State University (1961), 139–41.

11. "Testimony of James L. Mooney," *Violations of Free Speech and Rights of Labor: Hearings Before a Subcommittee of the Committee on Education and Labor, United States Senate . . . Part 14, The Chicago Memorial Day Incident*, June 30, July 1 and 2, 75th Cong., 1st Sess. (Washington, DC: US Government Printing Office, 1937), 4684 [hereafter cited as *LSC Part 14*]; "Testimony of John Riffe," *LSC Part 14*, 4865.

12. "Testimony of John Riffe," *LSC Part 14*, 4865.
13. "Testimony of Gus Yuratovac," *LSC Part 14*, 4874–75.
14. "Testimony of George A. Patterson," *LSC Part 14*, 4879–80; Patterson autobiography, bk. 2, p. 16, box 9, Patterson Papers.
15. "Testimony of George A. Patterson," *LSC Part 14*, 4879–80.
16. Ibid., 4881.
17. "Violence Flares as 25,000 Walk Out: Forty Arrested; Plants Threaten to Close," *Chicago Herald and Examiner*, May 27, 1937.
18. George Patterson interview, Dec. 1970–Jan. 1971, 80, Elizabeth Balanoff Labor Oral History Collection, Roosevelt University, Chicago.
19. "Testimony of Gus Yuratovac," *LSC Part 14*, 4875.
20. Ibid., 4876.
21. "Steel Strikers Battle Guards at Indiana Mill: News of Strife Fans the Temper of Workers at South Chicago," *Chicago Daily News*, May 29, 1937; "Testimony of George Patterson," *LSC Part 14*, 4884–85.
22. "Testimony of Thomas Kilroy," *LSC Part 14*, 4727.
23. "Testimony of George Patterson," *LSC Part 14*, 4885–86; "Report No. 46, Memorial Day Incident," *LSC Part 14*, 6; "23 Hurt in So. Chicago Steel Strike Riot: Police Clubs and Pistol Fire Turn Back March on Plant; Rout 1,500 C.I.O Paraders," *Chicago Herald and Examiner*, May 29, 1937.

CHAPTER THREE: Memorial Day 1937

1. Patterson autobiography, bk. 2, pp. 20–21, box 9, Patterson Papers; Howard Fast, "An Occurrence at Republic Steel," in Isabel Leighton (ed.), *The Aspirin Age, 1919–1941* (New York: Simon and Shuster, 1949), 384; Meyer Levin, *Citizens: A Novel* (New York: Viking Press, 1940), 11–12; Nathan Godfried, *WCFL: Chicago's Voice of Labor, 1926–1978* (Urbana: University of Illinois Press, 1997), 181–82; Denning, *Cultural Front*, 287.
2. Patterson autobiography, bk. 2, pp. 23–24, box 9, Patterson Papers.
3. Mollie West, interview by Araceli Ramirez and Sherry Oliphant for George Washington High School, Feb. 1995, in the author's possession.
4. "Testimony of Lupe Marshall," *LSC Part 14*, 4946;"Testimony of George R. Higgins," *LSC Part 14*, 4813; "Testimony of James L. Mooney," *LSC Part 14*, 4707–8; "Testimony of Jacob C. Woods," *LSC Part 14*, 4791–92; "Testimony of Lawrence J. Lyons," *LSC Part 14*, 4771–73.
5. "Testimony of Meyer Levin," *LSC Part 14*, 4894; "Testimony of John Lotito," *LSC Part 14*, 4938; "Testimony of Harry N. Harper," *LSC Part 14*, 4961; "Testimony of James Stewart," *LSC Part 14*, 4910; "Testimony of Anton Goldasic," *LSC Part 14*, 4935; "Testimony of Max Guzman," *LSC Part 14*, 4942.
6. "Testimony of Lupe Marshall," *LSC Part 14*, 4947–48; "Testimony of James Stewart," *LSC Part 14*, 4912.

7. "Testimony of Frank W. McCullogh," *LSC Part 14*, 4904; "Testimony of Chester B. Fisk," *LSC Part 14*, 4898; "Testimony of Harry N. Harper," *LSC Part 14*, 4961.

8. "Testimony of Harry N. Harper," *LSC Part 14*, 4961; "Testimony of John Lotito," *LSC Part 14*, 4939.

9. Patterson autobiography, bk. 2, p. 24, Patterson Papers; "Affidavits of Witnesses—Exhibit 3409: Marilee Kone Statement," *LSC Part 15-D*, 6739; "Testimony of Chester B. Fisk," *LSC Part 14*, 4898; "Testimony of Lupe Marshall," *LSC Part 14*, 4948; West, interview.

10. "Affidavits of Witnesses—Exhibit 3409: Marilee Kone Statement," *LSC Part 15-D*, 6739; "Testimony of Chester B. Fisk," *LSC Part 14*, 4898; "Testimony of Lupe Marshall," *LSC Part 14*, 4948; "Affidavits of Witnesses—Exhibit 3408: J. Gordon Bennett Statement," *LSC Part 15-D*, 6738.

11. "Testimony of Harry N. Harper," *LSC Part 14*, 4962.

12. "Testimony of Lupe Marshall," *LSC Part 14*, 4950–51.

13. Ibid.; "500 Policemen Called to Halt Strike March: Reinforcements Ready as Pickets Near S. Chicago Plant," *Chicago Daily News*, Jun. 1, 1937.

14. "Affidavits of Witnesses—Exhibit 3416: Louis Calvano Statement, July 1, 1937," *LSC Part 15-D*, 6747.

15. "Testimony of Chester B. Fisk," *LSC Part 14*, 4899.

16. "Riot Eyewitness Describes Fight in Which 9 Were Killed," *Chicago Daily News*, Jun. 18, 1937; Sean Callahan, "Out of the Past: A Striking Picture from Chicago's History Is Preserved on Film," *Daily Southdown*, Dec. 19, 1997, in Southeast Chicago Historical Project Collection, Calumet Park Field House, Chicago [hereafter abbreviated as SCHPC].

17. "Testimony of Archibald G. Paterson," *LSC Part 14*, 4966; Meyer Levin, "Slaughter in Chicago," *Nation* 144 (Jun. 12, 1937): 671; "Testimony of Meyer Levin," *LSC Part 14*, 4894–95.

18. Patterson autobiography, bk. 2, p. 25, Patterson Papers.

19. "Eyewitness Tells Horrors of Fighting: Housewife Averts Eyes from Sight; Bystander Rescues Boy of 9, Wounded in Foot," *Chicago Herald and Examiner*, May 31, 1937.

20. "Appendix, 1439: Harry Harper Statement, June 1937," *LSC Part 14*, 5061.

21. "Testimony of Lupe Marshall," *LSC Part 14*, 4951.

22. Ibid., 4952.

23. "Testimony of Archibald G. Paterson," *LSC Part 14*, 4967–68; "Affidavits of Witnesses—Exhibit 3414: John Jablonski Statement, June 29, 1937," *LSC Part 15-D*, 6746.

24. Paul Y. Anderson, "Suppressed Film Reveals Police as Aggressors in Strike Riot," *Washington Post*, Jun. 17, 1937. Also quoted in Sidney Lens, *The Labor Wars: From the Molly Maguires to the Sitdowns* (New York: Doubleday, 1973), 320.

25. "Testimony of Chester B. Fisk," *LSC Part 14*, 4900.

26. "Appendix, 1439: Louis Selenik Statement, June 24th, 1937," *LSC Part 14*, 5070.

27. "Testimony of Dr. Lawrence Jacques," *LSC Part 14*, 4984–86.
28. Ibid., 4986; Frank John Fonsino, "An Oral History Version of the Memorial Day Massacre at Republic Steel," 3, 13–14, SCHPC.
29. "Affidavits of Witnesses—Exhibit 3409: Marilee Kone Statement," *LSC Part 15-D*, 6739–40; "Affidavits of Witnesses—Exhibit 3413: James C. Row Statement," *LSC Part 15-D*, 6745.
30. "Testimony of Dr. Lawrence Jacques," *LSC Part 14*, 4986–87; "Affidavits of Witnesses—Exhibit 3413: James C. Row Statement," *LSC Part 15-D*, 6745.
31. "Exhibit 1426: Harry Harper Statement," *LSC Part 14*, 5061–62; "Testimony of Harry N. Harper," *LSC Part 14*, 4964.
32. "Testimony of Lupe Marshall," *LSC Part 14*, 4952–53.
33. Ibid., 4954–56.
34. "Appendix, 1439: Harry Harper Statement, June 1937," *LSC Part 14*, 5063.
35. "Testimony of Edwin J. Kennedy," *LSC Part 15-D*, 6917–18.
36. "Testimony of Lawrence J. Lyons," *LSC Part 14*, 4757.
37. "Exhibit 1380: Testimony of William H. Cannon, June 29, 1937," *LSC Part 14*, 5039; "Testimony of James L. Mooney," *LSC Part 14*, 4707.
38. "Exhibit 1361: Officer George Higgins Statement, June 28, 1937," *LSC Part 14*, 5029.
39. Ibid.
40. "Testimony of Dr. Lawrence Jacques," *LSC Part 14*, 4992–93.
41. Ibid., 4990–94; Fonsino, "Oral History Version of the Memorial Day Massacre," 12; William Hal Bork, "The Memorial Day 'Massacre' of 1937 and Its Significance in the Unionization of the Republic Steel Corporation," master's thesis, University of Illinois at Urbana-Champaign (1975), 110–17.
42. "Testimony of Dr. Lawrence Jacques," *LSC Part 14*, 4994–98; "Testimony of Lupe Marshall," *LSC Part 14*, 4954.
43. Alice Hoffman interview with Les Thornton, Nov. 12, 1970, 15, Oral History Projects, Department of Labor Studies, Paterno Library, Pennsylvania State University.

CHAPTER FOUR: Red Scare and Popular Resistance

1. Quoted in James Green, *Death in the Haymarket: A Story of Chicago, the First Labor Movement, and the Bombing That Divided Gilded Age America* (New York: Pantheon Books, 2006), 270.
2. Quoted in Philip Dray, *There Is Power in a Union: The Epic Story of Labor in America* (New York: Doubleday, 2010), 155.
3. Milton Howard, "Chicago: Testing Ground for American Fascism; Slaying of Five Steel Workers Another Act of Labor Suppression in Record of Kelly-Courtney-Nash Reign," *Daily Worker*, Jun. 1, 1937; Kim Phillips-Fein, *Invisible Hands: The Businessmen's Crusade Against the New Deal* (New York: W. W. Norton, 2009), 13–24.
4. "Pin Steel Riot on Red Agents: 6th Victim Dies, 65 Prisoners Face Court To-

day," *Chicago Daily Tribune*, Jun. 2, 1937; "How Trotzky's Revolt Won: A Warning to U.S.; Analyze Red Minority's Subtle Weapons," *Chicago Daily Tribune*, Jun. 2, 1937.

5. "Exhibit 1332: Make Mills Report on 'Steel Strike Riot,' June 2, 1937," *LSC Part 14*, 5009; "Exhibit 1333: Make Mills Report on 'Steel Strike Riot,' June 16, 1937," *LSC Part 14*, 5010; "Affidavits of Witnesses—Exhibit 3525: Ada Leder Statement," *LSC Part 15-D*, 6869.

6. "Steel Strike Peace Conference Fails: 5 Dead, 4 More Dying after Riot," *Daily Calumet*, Jun. 1, 1937; "Pin Steel Riot on Red Agents."

7. "Roosevelt Uses Any Weapon to Fight Business: Aids Say 'You Ain't Seen Nothing Yet,'" *Chicago Daily Tribune*, Jun. 7, 1937.

8. "Thousands Attend CIO Rally: 20,000 Hear CIO Pledge Finish Fight," *Chicago Evening American*, Jun. 18, 1937.

9. "Horner Intervenes in Chicago Strike: Fifth Death in Riot; Governor Confers with Union, Steel and City Officials and Police in Peace Move," *New York Times*, Jun. 1, 1937; "Steel Pickets Invade Loop," *Chicago Evening American*, Jun. 1, 1937; "500 Policemen Called to Halt Strike March: Reinforcements Ready as Pickets Near S. Chicago Plant," *Chicago Daily News*, Jun. 1, 1937; "Police Fear New Violence in Mills Area: 948 Massed," *Chicago Herald and Examiner*, Jun. 2, 1937, 2. Also see "Strike Tussle in Loop: Woman Battles Bystander," *Chicago Herald and Examiner*, Jun. 2, 1937, 2.

10. "5,000 Strikers Protest Police Action," *Chicago Herald and Examiner*, Jun. 1, 1937; "5,000 Join 'Hero' Parade: Bittner at Mass Meeting Demands Indictment of Police on Murder Charges," *New York Times*, Jun. 1, 1937.

11. Albert W. Palmer, "An Apology to the Dead: Report of a Talk at Citizens Mass Meeting, Tuesday Evening, June 8, 1937, Chicago Civic Opera House," box 37, folder titled "Citizens' Joint Commission of Inquiry on South Chicago Memorial Day Incident (Republic Steel Riots) 1937," Graham Taylor Collection, Newberry Library, Chicago.

12. "Riot Quiz Demanded as 4,600 Boo Police," *Chicago Evening American*, Jun. 9, 1937.

13. Ibid.; Palmer, "An Apology to the Dead."

14. "Riot Quiz Demanded as 4,600 Boo Police"; Robert Morss Lovett, *All Our Years: The Autobiography of Robert Morss Lovett* (New York: Viking Press, 1948), 263.

15. Palmer, "Apology to the Dead," 13.

16. Sofchalk, "Little Steel Strike of 1937," 337.

17. Joseph M. Turrini, "The Newton Steel Strike: A Watershed in the CIO's Failure to Organize 'Little Steel,'" *Labor History* 38 (Spring/Summer 1997): 258–60.

18. Quoted in Bernstein, *Turbulent Years*, 494, 496.

19. On July 24, at the apex of the Little Steel Strike, Philip Murray sent a circular to all SWOC lodges and members, reminding them of the primacy of their signed contracts. "Strikes, walkouts, and other stoppages of work constitute a violation both of our contracts and the policy of the Steel Workers Organizing Committee. Under no circumstances should such acts take place." Murray called on members to exert their "good judgment, common sense, and solidarity" by respecting their contracts.

What mattered most to Murray was the "integrity of our contracts," not the possibility that coordinated mass walkouts could bring the industry to a grinding halt. See Philip Murray, "Official Circular to All Staff Members, Local Lodge Officers and Members," Jul. 24, 1937, box 5, folder titled "Steel Workers Organizing Committee, SWOC Bulletins, 1937–38," Howard Curtiss Papers, HCLA.

20. "Observers' Report to Commanding General District of Northern Illinois," Jun. 26, 1937, box 411, folder 1, Governor Henry Horner Papers, Abraham Lincoln Presidential Library and Museum, Springfield, Illinois.

21. Murray, "Official Circular," Howard Curtiss Papers.

22. Patterson autobiography, bk. 2, pp. 107–9, Patterson Papers.

23. Ibid., 112.

24. On the class character of the labor struggles in the 1930s, which routinely spread beyond a given employer to involve workers in an entire industry or city, see Joe Burns, *Reviving the Strike: How Working People Can Regain Power and Transform America* (Brooklyn, NY: Ig Publishing, 2011), 119.

25. Roger Biles, "Edward J. Kelly: New Deal Machine Builder," in Green and Holli, *The Mayors*, 111–15.

26. Ibid., 113–20; Roger Biles, *Big City Boss in Depression and War: Mayor Edward J. Kelly of Chicago* (DeKalb: Northern Illinois University Press, 1984), 76–84; Robert A. Slayton, "Labor and Urban Politics: District 31, Steel Workers Organizing Committee, and the Chicago Machine," *Journal of Urban History* 23 (Nov. 1996): 35–45.

27. "16,000 Attend Steel Union's Mass Meeting: Speakers Declare C.I.O. Will Not Rest Until Companies Sign," *Chicago Daily News*, Jun. 18, 1937.

28. "Testimony of John C. Prendergast," *LSC Part 14*, 4672–73.

29."Chicago C.P. Urges City-Wide Protest," *Daily Worker*, May 31, 1937; "948 Police Mobilized at Mills: Riot Victim a Red, Quiz Told; 25,000 to Attend Mass Burial," *Chicago Herald and Examiner*, Jun. 2, 1937; on SWOC member Joe Germano's theory that the communists engineered the Memorial Day incident in order to produce martyrs, see "Oral History Interview #2 with Joe Germano," Jun. 21, 1972, 2–5, Oral History Projects, Department of Labor Studies, Special Collections, Paterno Library, Pennsylvania State University.

30. "Testimony of James L. Mooney," *LSC Part 14*, 4701–2; "Testimony of Thomas Kilroy," *LSC Part 14*, 4746–47; "Testimony of John C. Prendergast," *LSC Part 14*, 4676–77.

31. "Testimony of James P. Allman," *LSC Part 14*, 4664; "Testimony of James L. Mooney," *LSC Part 14*, 4689–90, 4707.

32. Donald Sofchalk, "The Chicago Memorial Day Incident: An Episode of Mass Action," *Labor History* 6 (Winter 1965): 36–37; La Follette quote from "Testimony of James L. Mooney," *LSC Part 14*, 4719; "Testimony of Thomas Kilroy," *LSC Part 14*, 4738–39.

33. "Testimony of Philip Igoe," *LSC Part 14*, 4821; "Testimony of Lawrence J. Lyons," *LSC Part 14*, 4760; "Testimony of George R. Higgins," *LSC Part 14*, 4813–14.

34. "Exhibit 1361: Memo of Interview, June 26—Chicago Shootings," *LSC Part 14*, 5029.

35. "Police Faced with Photos of Beatings: Chicago Official Admits Scene in Picture Is 'Brutal,'" *Washington Post*, Jul. 1, 1937; "Chicago's Brutal Police," *Washington Post*, Jul. 2, 1937.

36. "Testimony of James L. Mooney," *LSC Part 14*, 4692–94.

37. "Testimony of Lawrence J. Lyons," *LSC Part 14*, 4768; "Testimony of Jacob C. Woods," *LSC Part 14*, 4793; "Exhibit 1361: Memo of Interview, June 26—Chicago Shootings," *LSC Part 14*, 5029.

38. "Exhibit 1627-C: Make Mills, Industrial Detail, to Commissioner of Police, June 14, 1937," *LSC Part 14*, 5160–62.

39. "Testimony of Lawrence J. Lyons," *LSC Part 14*, 4769.

40. "Exhibit 1361: Memo of Interview, June 26—Chicago Shootings," *LSC Part 14*, 5029.

41. "Testimony of James L. Mooney," *LSC Part 14*, 4692.

42. Vorse, *Labor's New Millions*, 131–32.

43. Mary Heaton Vorse, "The Tories Attack through Steel," *New Republic*, Jul. 7, 1937, 246.

44. Howard, "Chicago: Testing Ground for American Fascism." On the origins of the assault on the New Deal in the 1930s that would eventually see business ideology vindicated, see Phillips-Fein, *Invisible Hands*, 3–25.

45. "Rep. Maverick Defends C.I.O. in Fiery Debate: Texan Hits Union Critics; Bitter Speeches Force Adjournment," *Washington Post*, Jul. 3, 1937.

46. Ibid.

47. Kennedy, *Freedom from Fear*, 349.

48. "Report No. 46, Part 2: Report on the Memorial Day Incident," *LSC Part 14*, 4–21.

49. Ibid., 34–35.

50. Ibid., 38–40.

51. "Report of the Citizens' Joint Commission of Inquiry," 3–8.

52. "Chicago Riot Films Stun Audience Here: Paramount Releases Pictures Taken during Memorial Day Steel Trouble," *New York Times*, Jul. 3, 1937.

53. Carol Quirke, "Reframing Chicago's Memorial Day Massacre, May 30, 1937," *American Quarterly* 60 (Mar. 2008), 149.

54. "Chicago Police Slugged, Kicked, Cursed Her, Trampled Wounded Strikers, Woman Charges," *Washington Post*, Jun. 29, 1937; "Labor's Gethsemane," *Washington Post*, Jul. 4, 1937; "Riot Film Backed by New Witnesses: Pictures from Suppressed Film Introduced at Strike Investigation," *New York Times*, Jul. 3, 1937; Jerold Auerbach, *Labor and Liberty: The La Follette Committee and the New Deal* (New York: Bobbs-Merrill, 1966), 127–28.

55. Paul Y. Anderson, "Suppressed Film Reveals Police as Aggressors in Strike Riot," *Washington Post*, Jun. 17, 1937; Lillian Elkin, "Paul Y. Anderson: The *Nation*'s Angry Man," *Nation* 200 (May 31, 1965): 584; Auerbach, *Labor and Liberty*, 127.

56. "Labor's Gethsemane."
57. Sofchalk, "Chicago Memorial Day Incident," 37–38.
58. Nelson Lichtenstein, "Politicized Unions and the New Deal Model: Labor, Business, and Taft-Hartley," in Sidney Milkis and Jerome Mileur (eds.), *The New Deal and the Triumph of Liberalism* (Amherst: University of Massachusetts Press, 2002), 144.
59. Kennedy, *Freedom from Fear*, 350–55; James Green, *The World of the Worker: Labor in Twentieth-Century America* (New York: Hill and Wang, 1980), 165.
60. Sofchalk, "Chicago Memorial Day Incident," 32–35.
61. Green, *World of the Worker*, 157.
62. Sofchalk, "Little Steel Strike," 121–23.
63. "Troops Guard Ohio Workers: Johnstown Martial Law Lifted; Girdler Calls Murray a 'Liar,'" *Washington Post*, Jun. 25, 1937.
64. Sofchalk, "Chicago Memorial Day Incident," 32, 38–39.
65. "Propaganda for Insurrection," *Chicago Daily Tribune*, Jul. 25, 1937.
66. "The Industrial War," *Fortune* 16 (Nov. 1937): 184.

CHAPTER FIVE: Little Steel and Class Warfare

1. Eric Shickler and Devin Caughey, "Public Opinion, Organized Labor, and the Limits of New Deal Liberalism, 1936–1945," *Studies in American Political Development* 25 (Oct. 2011): 170–73.
2. Harry J. Brevis, "Radio and the Steel Strike," *Nation* 145 (Jul. 10, 1937): 55.
3. "Law and Order," *Nation* 145 (Jul. 3, 1937): 5.
4. Ibid.
5. McKay, *Rebels, Reds, and Radicals*, 141.
6. Quoted in Ahmed White, "The 'Little Steel' Strike of 1937: Class Violence, Law, and the End of the New Deal," Aug. 2010, http://works.bepress.com/cgi/viewcontent.cgi?article=1001&context=ahmed_white/.
7. "Testimony of John C. Prendergast," *LSC Part 14*, 4672–73; "948 Police Mobilized at Mills"; Bernstein, *Turbulent Years*, 498.
8. "Radio Address by Samuel C. Evett, District Secretary, Steel Workers Organizing Committee, W.I.N.D., Gary, Indiana," Apr. 26, 1939, box 1, folder 5, CRA [Calumet Regional Archives] 137, Evett Papers.
9. "Law and Order," 6.
10. "Public Opinion on Labor," *LSC Part 15-D*, 8032–33.
11. "Public Opinion on Labor Unions Splits Along Class Lines," *Washington Post*, Jul. 4, 1937.
12. Sofchalk, "Little Steel Strike of 1937," 271–77; Sofchalk, "Chicago Memorial Day Incident," 26–29; Michael Speer, "The Little Steel Strike: Conflict for Control," *Ohio History* 78 (1969): 279–82; Len De Caux, *Labor Radical: From the Wobblies to CIO* (Boston: Beacon Press, 1970), 291–93; Joseph Ator, "Our Chief Peril: A Blind Faith in Majorities," *Chicago Daily Tribune*, Jun. 4, 1937; Ator, "How Dictators Set

the Stage: Is U.S. on Road? Iron Rule Has Its Roots in Class Hatred," *Chicago Daily Tribune*, Jun. 3, 1937.

13. Sofchalk, "Little Steel Strike of 1937," 387–88; "United Packinghouse Workers Industrial Union Resolution, Armour Branch," n.d., box 411, folder 1, Governor Henry Horner Papers.

14. Ellen Schrecker, *Many Are the Crimes: McCarthyism in America* (Princeton, NJ: Princeton University Press, 1998), 64–85, quote on 69.

15. Ibid., 123–24, 130–31.

16. Zieger, *CIO, 1935–1955*, 141–47, 169–82.

17. Bruce Nelson, "'Give Us Roosevelt': Workers and the New Deal Coalition," *History Today* 40 (Jan. 1990): 2–9.

18. Benjamin Stolberg, "Big Steel, Little Steel, and C.I.O.," *Nation* 145 (Jul. 24, 1937): 121–22.

19. John L. Lewis, "Report of President John L. Lewis of the CIO," proceedings of the second constitutional convention of the CIO, "Speeches and Reports," John L. Lewis papers, Part 1, reel 2, (microfilm). The ideological contest following the Memorial Day Massacre was largely over the meaning of law. The *Chicago Daily News* editorialized that the "C.I.O., by its encouragement of lawlessness in sit-down strikes, lost public sympathy. Its cause has not been helped by the march on the city police at South Chicago." See "The Riot at South Chicago," *Chicago Daily News*, Jun. 2, 1937. Variations on this theme of CIO lawlessness would appear in Chicago's major dailies and in several newspapers throughout the country.

20. Michael Goldfield, "Worker Insurgency, Radical Organization, and New Deal Labor Legislation," *American Political Science Review* 83 (Dec. 1989): 1269–70.

21. Roger Keeran, "The International Workers Order and the Origins of the CIO," *Labor History* 30 (Summer 1989): 396.

EPILOGUE: Rethinking the Massacre

1. Feurer, *Radical Unionism in the Midwest*, xvii, 236.

2. Burns, *Reviving the Strike*, 14–25, 29–34.

3. Ibid., 20–31.

4. Edward Levinson, "Labor on the March," *Harper's Magazine* 174 (May 1937): 646.

5. Peter Zwiebach, "An American Tragedy: The Decline of U.S. Unionism and its Human Rights Implications," *Human Rights and Welfare* 5 (2005): 101–2, www.du.edu/korbel/hrhw/volumes/2005/zwiebach-2005.pdf.

6. James Gray Pope, Peter Kellman, and Ed Bruno, "The Employee Free Choice Act and a Long-Term Strategy for Winning Workers' Rights," *Working USA: The Journal of Labor and Society* 11 (Mar. 2008): 136.

SUGGESTED FURTHER READING

The study of the Memorial Day Massacre invariably revolves around the study of the labor movement and the industry that produced it. One of the earliest and most important treatments of the labor uprising that produced the unionization drive in the steel industry is Irving Bernstein's *The Turbulent Years* (New York: Houghton Mifflin, 1970). Bernstein has offered a comprehensive treatment of the labor uprising of the 1930s, albeit one that emphasizes the maneuvering of union leaders and political elites. There are dozens of studies of militant labor activism and the place of the Communist Party in the upsurge of union activity in the 1930s, but Bruce Nelson's *Workers on the Waterfront: Seamen, Longshoremen, and Unionism in the 1930s* (Urbana: University of Illinois Press, 1990) probably best captures the ideological dynamism and social ferment of that era. The vital importance of New Deal liberalism in the effort to achieve workers' rights and social justice in the 1930s is explored in Jerold Auerbach's pivotal *Labor and Liberty: The La Follette Committee and the New Deal* (New York: Bobbs-Merrill, 1966). Auerbach has highlighted the indispensable contribution of New Deal liberals, such as Robert La Follette and Robert Wagner, to the campaign to expose the violent and sinister tactics used by company after company to thwart the freedoms of speech and assembly by workers trying to achieve democracy in industry.

Any exploration of the movement to organize steelworkers necessarily involves a study of the steel industry in which they worked. Some of the most important examinations of the steel industry and American industrialization, the impact of steel manufacturing in a particular location over time, and the drive for labor and civil rights are Thomas Misa's *A Nation of Steel: The Making of Modern America, 1865–1925* (Baltimore: Johns Hopkins University Press, 1995); John Hinshaw's *Steel and Steelworkers: Race and Class Struggle in Twentieth-Century Pittsburgh* (Albany: State University Press of New York, 2002); and Mark Reutter's *Making Steel: Sparrows Point and the Rise and Ruin of American Industrial Might* (Urbana: University of Illinois Press, 2005). Unique within this genre of works on the social impact of the steel industry is Jack Metzgar's *Striking Steel: Solidarity Remembered* (Philadelphia: Temple University Press, 2000), which sensitively depicts the contribution of labor unions to the creation of postwar prosperity and economic security while deftly portraying working-class culture in urban America.

The dynamic relationship between working-class activism and New Deal liberalism is most fully developed in David Brody's *Workers in Industrial America: Essays on* the Twentieth Century Struggle (New York: Oxford University Press, 1980), a book that also emphasizes the "job conscious" character of American workers who sought industrial decency rather than workers' control at the point of production. For Brody, the 1930s and 40s saw the emergence of a system of "workplace contractualism," grounded in New Deal law, that represented a monumental improvement over the virulent repression of the 1920s. He has elaborated on the obstacles as well as the possibilities facing industrial unionism in steel companies in the 1930s in "The Origins of Modern Steel Unionism: The SWOC Era," an essay appearing in Paul F. Clark, Peter Gottlieb, and David Kennedy (eds.), Forging a Union of Steel: Philip Murray, SWOC, and the United Steelworkers (Ithaca, NY: ILR Press, 1987). Brody has also advanced the study of how the Employee Representation Plans, or company unions, became the embryo of authentically independent unionism, an effort which has been amplified significantly by James Rose in Duquesne and the Rise of Steel Unionism (Urbana: University of Illinois Press, 2001). Rosemary Feurer's Radical Unionism in the Midwest, 1900–1950 (Urbana: University of Illinois Press, 2006) is an exhilarating and important exploration of democratic unionism in the New Deal and postwar eras. In contrast, Nelson Lichtenstein's Labor's War at Home: The CIO in World War II (New York: Cambridge University Press, 1982; new edition, Philadelphia: Temple University Press, 2003) is a sobering and provocative examination of how political elites, union leaders, and corporate executives used wartime mobilization to control labor activism.

Equally important for understanding the cultural and ideological forces that produced the labor upheaval of the 1930s are Lizabeth Cohen's Making a New Deal: Industrial Workers in Chicago, 1919–1939 (New York: Cambridge University Press, 1990), and Rick Halpern's Down on the Killing Floor: Black and White Workers in Chicago's Packinghouses, 1904–1954 (Urbana: University of Illinois Press, 1997). Cohen's book lucidly explores the relationship between the unions' working-class formation, labor union activism, and the social democratic reforms sponsored by the Roosevelt administration. For Cohen, the fostering of a working-class "culture of cohesion" was indispensable to the achievements of the CIO in the 1930s. Halpern's book examines the interaction of race and class in Chicago's iconic meatpacking industry. Down on the Killing Floor elucidates the development of one of the most progressive unions in the New Deal era and its relationship to the broader movement for industrial democracy. Robert Slayton's "Labor and Urban Politics," Journal of Urban History 23 (Nov. 1996): 29–65, is an illuminating article that explores how the Chicago political machine absorbed and contained the once-militant Steel Workers' Organizing Committee.

Any study of the Memorial Day incident in Chicago is a byproduct and subset of the study of American labor unionism, the New Deal, and the working-class activism of the 1930s. For the most part, examinations of the massacre and the strike that produced it follow the pattern of labor history itself, with the earliest treatments emphasizing the effort to achieve the protection of the law in an industry defined by despotism and antiunionism. The most comprehensive and still thought-provoking treatment of the Little Steel Strike is Donald Sofchalk's PhD dissertation, "The Little Steel Strike of 1937" (Ohio State University, 1961). Sofchalk has sympathetically presented the case for union recognition, but he has also acknowledged the weakness of SWOC support during the strike effort at Republic Steel's Burley Avenue plant in 1937. His subsequent publication, the seminal "The Chicago Memorial Day Incident: An Episode of Mass Action," *Labor History* 6 (Winter 1965): 3–43, emphasizes the legitimacy of SWOC's position and the brutality of the response by the police, but in it Sofchalk has also concluded that what "had promised to be a successful display of militant solidarity turned out to be a costly mistake" (36). Since then, a number of historians have expanded the study of the virulently antiunion tactics of Republic Steel and its president, Tom Girdler, as well as the class character of the Little Steel Strike. As examples, see Louis Leotta, "Girdler's Republic: A Study in Industrial Warfare," *Cithara* 11 (1971): 41–66; Daniel J. Leab, "The Memorial Day Massacre," *Midcontinent American Studies Journal* 8 (1967): 3–17; and Michael Speers, "The Little Steel Strike: Conflict for Control," *Ohio History* 78 (1969): 273–87.

More recently, Zaragosa Vargas's *Labor Rights Are Civil Rights: Mexican American Workers in Twentieth-Century America* (Princeton, NJ: Princeton University Press, 2005) highlights the Latino contribution to the movement for industrial unionism in the steel industry. Vargas's book also restores the inimitable Lupe Marshall to our historical consciousness. Vargas's important study was soon followed by Carol Quirke's imaginative and insightful "Reframing Chicago's Memorial Day Massacre, May 30, 1937," *American Quarterly* 60 (Mar. 2008): 129–57. In her more recent *Eyes on Labor: News Photography and America's Working Class* (New York: Oxford University Press, 2012), Quirke has focused on the representation and politicization of the revelatory newsreel footage shot by Orlando Lippert. Mainstream media has played a decisive and often negative role in reporting on organized labor, and Quirke's essay is an important contribution to our understanding of the lenses that have been used to filter our perception of American labor activism. For a study that emphasizes the place of the Memorial Day Massacre in the broader effort toward industrial democracy, see Michael Dennis, *The Memorial Day Massacre and the Movement for Industrial Democracy* (New York: Palgrave Macmillan, 2010). His "Chicago and the Little Steel Strike," *Labor History* 53 (May 2012): 167–205, narrows that perspective somewhat and situates the Little Steel Strike at the center of a movement for social democracy and social justice in the 1930s. In "The Little Steel Strike of 1937: Class Violence, Law, and the End of the New Deal" (Aug. 2010), http://works.bepress.com/cgi/viewcontent.cgi?article=1001&context=ahmed_white/, Ahmed White has argued that the strike was a pivotal moment in the class confrontation of the 1930s, since it accelerated the CIO's movement away from progressive unionism while graphically illustrating the Roosevelt administration's unwillingness to counter the prerogatives of capitalism.

For a longer-range view of the American labor movement, one that address its past decline and its prospects for workers today, see Nelson Lichtenstein's *State of the Union: A Century of American Labor* (Princeton, NJ: Princeton University Press,

2002). To understand the key role that strikes, as well as their associated tactics of fostering class solidarity, have played in permitting at least some American workers to achieve a decent standard of living in the twentieth century, see Joe Burns, *Reviving the Strike: How Working People Can Regain Power and Transform America* (Brooklyn, NY: Ig Publishing, 2011). Burns also discusses the vitally important question of law, which has steadily been transformed into one of the most effective antilabor weapons in modern America. Key additions to this body of literature include James Gray Pope, "How American Workers Lost the Right to Strike, and Other Tales," *Michigan Law Review* 103 (Dec. 2004): 518–53; James Gray Pope, "Worker Lawmaking, Sit-Down Strikes, and the Shaping of American Industrial Relations, 1935–1958," *Law and History Review* 24 (Spring 2006): 45–113; and James Gray Pope, Peter Kellman, and Ed Bruno, "The Employee Free Choice Act and a Long-Term Strategy for Winning Workers' Rights," *Working USA: The Journal of Labor and Society* 11 (Mar. 2008): 125–44.

Finally, for a consideration of why workers join unions at all—including the role of unions in promoting democratic representation and equality in the workplace, the times when they have failed to do so, and their prospects in an era of economic globalization—see Michael D. Yates, *Why Unions Matter* (New York: Monthly Review Press, 1998); Steve Early, *Embedded with Organized Labor: Journalistic Reflections on the Class War at Home* (New York: Monthly Review Press, 2009); and Kim Moody, *US Labor in Trouble and Transition: The Failure of Reform from Above, the Promise of Revival from Below* (New York: Verso, 2007).

INDEX

Page numbers in *italics* refer to figures and photographs.

African Americans in Communist Party, 11, 18. *See also* ethnicity
Allman, James, 78
American Federation of Labor (AFL), 7, 16, 76
Anderson, Hilding, 53, 59
Anderson, Paul Y., 50, 88
anticommunism, 84–85, 100–101. *See also* Red Scares
antilabor violence. *See* violence, antilabor

Badornac, Emil, 24
Beck, Ralph, 45–46, 78, 85
Bennett, J. Gordon, 43
Bethlehem Steel, 1
bigotry, ethnic and racial, 26–31, 81
Bittner, Van: contracts and, 20; at Indiana Harbor rally, 67; on Memorial Day, 37; on policy and issues, 97–98; reassurances by, 72, 77, 95; on Republic Steel, 69; on US Steel, 76
Bohrte, Uva and Melsina, 47
Brevis, Harry J., 95–96
Bridewell Hospital, 53–54
Browder, Earl, 64
Burns, Joe, 111
Burnside Hospital, 54

Calvano, Louis, 45, 59
Cannon, William H., 56
Carnegie, Andrew, 7
Causey, Alfred, 59, 88
Chicago: antilabor violence in, 11; New Deal in, 74, 75; political establishment in, 28, 30, 74–78, 103; Popular Front movement in, 36, 51, 68; race riot of 1919 in, 29; Southeast Side of, 38; strikes in, 27–28. *See also* Chicago police
Chicago Citizens' Rights Committee protest, 68–70, 87
Chicago Daily Tribune, 63–64, 66–67, 83, 89, 92–93
Chicago Department of Law, 25
Chicago police: after riot, 74–78; cordon of, 39–40; culture and attitudes of, 26–31; handling of injured persons by, 47–55, 79, 86; harassment by, 2, 25–26; infiltration by, 5; motives behind conduct of, 79–80; newsreel release and, 87–89; property rights and, 66; public opinion of, 92–93; Red Scare launched by, 62–67; speeches castigating, 2; as strikebreakers, 86–87; testimony of, 55–58, 78–79; Unemployment Day demonstration and, 11; as violating labor law, 34; violence by, vii, 3, 6, 85–88; weapons of, 78
Chicago Repertory Group, 36, 37
Chicago Tribune, 55
Childs, Morris, 77
CIO. *See* Congress of Industrial Organizations
civil disobedience: mass protests as, 111; nonviolent, 113; success of, 111–12
class conflict: evidence for in public opinion, 99–100; forces behind, 108–9; labor struggles and, 127n24; Memorial Day march as evidence of, 60–61; in 1930s, 3–4; Patterson and, 73–74; sit-down strikes and, 17; steel magnates and, 98–99; subculture of policing and, 26–27
Communist Party: in America, 10–12, 15; labor movement and, 26, 37; massacre and,

Communist Party (*cont.*)
    63–67; Popular Front movement and, 23;
    SWOC and, 17–18. *See also* anticommunism;
    Red Scares
Congress of Industrial Organizations (CIO):
    capitulation of, 106; during Cold War, 104;
    commitment to reform of, 102–3; general
    strike in Ohio, 70–71; Girdler on, 92; industrial unionism and, 16; as lawless, 130n19;
    Little Steel Strike and, 7; mass meeting of,
    72; *Nation* criticism of, 97; opposition to, 83,
    84–85; public opinion of, 94–96. *See also*
    Steel Workers Organizing Committee
Courtney, Thomas, 66
Cox, Eugene, 84
Crowe Name Plate Company, 26

Daily Worker, 63
Daniels, Henderson, 18
Davey, Martin L., 71, 96
Democratic Party, 8, 97, 103–4
dissent, suppression of, 112–13
Douglas, Paul, 68

Edwards, Frank, 18
Esposito, Dominick, 35
ethnicity: of Communist Party, 11, 18; of labor
    movement, 5, 23; of militant masses, 26–27;
    police bigotry and, 81; of police officers, 29
Evett, Samuel C., 98

Fair Labor Standards Act of 1938, 90
Fansteel Metallurgical Corporation, 17
Fisk, Chester B., 37, 41, 42, 45, 50–51
Fontecchio, Nick, 34–35, 37
Foster, William Z., 11
Francisco, Leo, 59

General Motors, 16, 18
Germano, Joe, 24, 101
Gill, Clayton, 68
Girdler, Tom, 19, 20, 31, 71, 92
Glaser, Paul, 26
Golden, Clinton, 20
Grace, Eugene, 72
Great Depression, 4–5, 9, 115, 117. *See also* New
    Deal

Guffey, Joseph, 92
Guzman, Max, 5, 51

Handley, Earl, 49–50, 58, 86, 88
Harper, Harry, 40, 41–42, 44, 47, 49–50,
    53–55, 76
"hate strikes," 105
Haymarket episode, 5, 27, 31, 62–63
Hickey, Joseph, 45, 53
Higgins, George, 5, 39, 40, 56–58, 79, 81
high school students, protests by, 4
Hillman, Sidney, 16
Hodes, Barnet, 25
Hoffman, Clare E., 84
Hoover, Herbert, 10
Howard, Earl Dean, 68

Illinois law on strikes and picketing, 25
Indiana Harbor, 59, 67, 72. *See also* Youngstown
    Sheet and Tube
industrial democracy: after tragedy, 67–74;
    movement for, 3–4, 63; opposition to, 85;
    themes of in plays, 36; wage and benefit
    improvements and, 102
industrial unionism, 16–21
International Workers Order, 18
investigations, requests for, 70

Jablonski, John, 49, 58
Jacques, Lawrence, 37, 51–52, 53, 59, 68–69
Johnson, Henry (Hank), 5, 37, 66
Johnson, Jayson, 51
Jones, Otis, 50, 59, 88
Jones & Laughlin decision, 19–20

Kamin, Alfred, 68
Kelly, Edward, 2, 34, 74–76, 103
Kennedy, Edwin J., 55
Kennedy, Thomas, 67
Kilroy, Thomas, 34, 40, 66, 78
Koch, Lucille, 35, 52
Kone, Marilee, 42, 52, 53
Krugar, Nick, 51
Kryczki, Leo, 2

labor movement: anticommunism and, 100–
    101; attempt to discredit, 63–67; catalysts

for, 12–14, 19; as counterbalance, 113–14; Democratic Party and, 103–4; elevation of, 72; human rights and, 4; interracial alliance of, 5, 23; legislators and, 15–16; *Nation* criticism of, 95–97; objective of, 112; opponents of, 79–81; participation in, 108; public opinion of, 89–92, 99–100; as public spectacle, 39; revival of, 21–23; social justice vision of, 104. *See also* Popular Front movement
La Follette, Robert, Jr., 78–79, 86
La Follette Committee hearings, 8, 78–87, 92–93, 99
Leder, Ada, 51, 65, 66, 88
Levin, Meyer, 37, 40, 68
Levinson, Edward, 113
Lewis, John L.: CIO and, 16; contracts and, 20; FDR and, 71; federal government and, 95; federal mediation and, 98; on men in steel industry, 106; radio address of, 22–23; role of, 33, 97; United Mine Workers and, 17; US Steel and, 18
Lippert, Orlando, 40, 87
Little Steel companies: overview of, 1–2; Roosevelt administration and, 81–82; Taylor-Lewis settlement and, 19; WWII and, 101–2. *See also* Little Steel Strike; *specific companies*
Little Steel Strike: CIO and, 7; as community revolt, 8–9; loss of, 101; media and, 95; participation in, 108–9; political meaning of, 73–74; Popular Front movement and, 36; property rights and, 66; reach of, 31; repercussions of, 89–93, 105–9; stakes in, 83–84; support for, 36–37. *See also* Memorial Day Massacre
Local 65 (SWOC), 39, 73
Los Angeles Times, 80
Lotito, John, 40, 42
Lovett, Robert Morss, 68, 70
Lyons, Lawrence, 40, 55–56, 81

MacNamara, John and James, 80
Marshall, Guadalupe: arrest of, 54; career of, 5, 29; eyewitness report of, 39, 40–41, 42, 44–45, 48–49; Higgins testimony about, 58, 79; labeled as communist, 65, 66; at march, 88

mass protests and demonstrations, 111. *See also* picketing; strikes
matrix events, 9, 60
Maverick, Maury, 84–85
McCarthyism, 101
McCullogh, Frank, 37, 41, 85
McKay, Ian, 9, 97
McKellar, Kenneth, 92
media: antilabor violence and, 90–91, 92–93; antiunionism and, 99, 100; discrediting of social protest by, 112–13, 114; Little Steel Strike and, 95–96. *See also specific newspapers*
Memorial Day Massacre: casualties of, 58–60; gunfire eruption at, 40–41, 42–44; injured marchers at, 47–54, 79, 86; interpretation and meaning of, vii, 60–61; Maverick on, 84; photographs of, 43, 57; public opinion of, 87–89; tear gas used at, 42–43, 43, 46; violence eruption at, 41–42, 44–46, 86. *See also* newsreels, Paramount
The Memorial Day Massacre and the Movement for Industrial Democracy (Dennis), vii–viii
middle class: activists from, 4–5, 36–37, 107, 108–9; Kelly and, 75; "Roosevelt recession" and, 90; views of organized labor of, 90–92, 99
Mills, Make, 30–31, 65, 81
Minors, Marjorie, 34
Mitckess, Ben, 35, 53
Mooney, James: communists and, 80; conspiracy theory and, 64–65; on language of marchers, 39–40; as police captain, 5; Riffe and, 31–32, 33; statements of, 42, 56, 78–79, 85; at Walker Vehicle Company strike, 26
Mrkonich, Virginia, 52
Murphy, Frank L., 96
Murray, Philip: Communist Party and, 17; contracts and, 20, 95; focus of movement and, 97, 98; public opinion and, 77; role of, 33; US Steel and, 72

National Industrial Recovery Act of 1933, 13–14
National Labor Relations (Wagner) Act, 14, 20, 92, 111, 112
National Labor Relations Board v. Jones & Laughlin Steel Corporation, 19–20

National War Labor Board, 101–2
Nation magazine, 95–97, 98–99
Nelson, Catherine, 88
Nelson, Steve, 12
neoliberalism, 114
New Deal: anticommunism and, 63–64, 100–101; in Chicago, 74, 75; labor and, 10–16; lost opportunities of, 105; opposition to, 82–84, 85; pace of reform, 107; skepticism about, 89–93
newsreels, Paramount: Anderson report after viewing, 50; of Causey, 59; event captured for, vii, 6; Movie Censorship Bureau and, 88–89; photograph from, 43; release of version of, 87
Norris-LaGuardia Act of 1932, 13–14

Oakes, Walter, 58, 79

Palmer, Albert W., 68, 69–70
Paramount newsreels. *See* newsreels, Paramount
Paterson, Archibald, 46, 49, 58–59
Patterson, Ben, 18
Patterson, Dorothy, 5, 21
Patterson, George: career of, 5; eyewitness report of, 46–47; as leader of march, 37, 39; on police, 33–34, 35, 42; to police, 35; on sit-down strikes, 17; SWOC and, 73–74
Pegler, Westbrook, 83, 89
picketing: altercations while, 5–6; injunctions to prevent, 13; as legal right, 2, 25, 34; at Republic Steel, 32, 33, 67
picnic of striking steelworkers, 1, 2
Polanyi, Karl, 8
police. *See* Chicago police
political establishment in Chicago, 28, 30, 74–78, 103
Popovich, Sam, 58
Popular Front movement: in Chicago, 36, 51, 68; connection between massacre and, 69–70; Communist Party and, 23, 66; community unionism and, 22; decline of, 100–104; limitations of, 70
Preis, Art, 18
Prendergast, John C., 31, 70, 77

private police systems, 8
progressive alliances, 110
Progressive Era, 10
protection rackets, 30

race riot of 1919 in Chicago, 29
racial composition: of Communist Party, 11, 18; of labor movement, 5, 23; of militant masses, 26–27; police bigotry and, 81; of police force, 29
Randolph, A. Philip, 68
Red Scares, 27, 62–67, 81, 100–101. *See also* Communist Party
Reed, Kenneth, 48–49, 58, 86
Reese, Jesse, 72
Republic Steel Corporation: antiunionism of, 20–21, 24–25; call for strike on, 31; Ohio plants of, 91, 92; SWOC and, 19–20; United Steelworkers of America and, 101; weapons of police and, 78. *See also* Girdler, Tom
Republic Steel Corporation, Burley Avenue plant: map of, 38; marches to, 3, 35; non-striking workers inside, 2–3, 34; picketing at, 32, 33, 67; sit-down strike at, 113; spontaneous sit-in at, 33–34; walkout at, 1
Riccio, Emil, 68
Riffe, John, 24, 31–32, 33, 37, 95
right-to-work states, 114–15
Roosevelt, Franklin D.: court-packing scheme of, 89–90; culpability of, 81–82; election of, 10; Kelly and, 74, 75–76; labeled as communist, 64, 67; labor and, 8; refusal to intervene by, 71, 97; worker devotion to, 107. *See also* New Deal
"Roosevelt recession," 90
Rothmund, Joseph, 48, 58, 59, 63–64, 66, 81
Rothmund, Margaret, 63, 64
Row, James C., 52–53

Sam's Place, 1, 2, 3, 37, 38, 52–53
Sandburg, Carl, 68
Sanford, Raymond, 37
Schrecker, Ellen, 101
Selenik, Louis, 51, 53
Senate Subcommittee on Education and Labor. *See* La Follette Committee hearings

Sharp, Malcolm P., 68
sit-down strikes, 16–17, 31–32, 91–92, 113
Slayton, Robert, 76
Smith, Gerald L. K., 95
Socialist Party, 12, 15
social justice movement and labor unions, 116
social protest, regeneration of, 10–12
social reform, discrediting of, 112–13
South Chicago Hospital, 53
Southeast Side of Chicago, 38
southern wing of anti-New Deal movement, 89–90
Spies, August, 63
steel industry: antilabor violence in, 33–34, 35, 82; as bastion of antiunionism, 7; crises in, 104; as form of dictatorship, 8; Lewis on men in, 106
Steel Workers Organizing Committee (SWOC): attacks on, 101; call for strike by, 31; cautiousness of, 94–95, 97–98; Communist Party and, 17–18; dependency on national government of, 98; failure to capitalize on indignation by, 76–78; failure to participate in rally and march, 37, 39; "Little Steel" companies and, 1–2; mass funeral sponsored by, 67; opposition to, 82–83; Republic Steel and, 19–20, 24; sit-down strikes and, 17, 31–32; US Steel and, 18; Women's Auxiliary, 5, 72. *See also* Murray, Philip
Stewart, James (Jim), 39, 40, 41, 73
Stolberg, Benjamin, 105, 106
strikes: in Chicago, 27–28; goals of, 8–9; "hate," 105; hiring of replacements during, 15; in Illinois, as legal, 25; injunctions to break, 13; sit-down, 16–17, 31–32, 91–92, 113; support for, 8; Wagner Act and, 14–15; wave of in 1937, vii, 16–17. *See also* Little Steel Strike; picketing
"supersedure," moment of, 97
SWOC. *See* Steel Workers Organizing Committee

Taft-Hartley Act, 90
Tagliori, Anthony, 53, 59
Taylor, Myron, 18
teachers, protests by, 4

Terkel, Studs, 37
Thomas, Elbert D., 80, 81, 86
Thornton, Les, 60–61
Tisdale, Lee, 45, 59

Unemployed Councils of the USA, 12–13, 27, 31
unemployment and Communist Party, 11–12
Unemployment Day demonstration, 11
unionism: history in debate over, 117; industrial, 16–21; opponents of, 114–15; proponents of, 115–17; stigmatizing of, 114
unionization: company resistance to, 1–2; democratization of society and, 108–9; goals of, 3–4; Wagner Act and, 14. *See also* labor movement; *specific unions*
United Auto Workers, 71
United Mine Workers, 17
United Packinghouse Workers, 100
United Steel Workers of America, 101–2, 103
US Steel: AFL and, 7; Carnegie-Illinois plant, 73; Carnegie-Illinois South Works Women's Auxiliary, 21–22; 1919 strike at, 27; SWOC and, 17, 18, 101; unionization of, 1

violence: American communists and, 66; by Chicago police, vii, 3, 6, 85–88; eruption of on Memorial Day, 41–42, 44–46, 86; Roosevelt administration implicated in, 64; suppression of dissent and, 112–13
violence, antilabor: belief in coercion and, 28; at Burley Avenue plant, 33–34, 35; in Chicago, 11; history of, 5–6, 82–83; media and, 90–91, 92–93. *See also* Memorial Day Massacre
Vorse, Mary Heaton, 4–5, 6, 19, 82–83, 112

wage equalization, as objective, 112
Wagner, Robert L., 14
Wagner (National Labor Relations) Act, 14, 20, 92, 111, 112
Walker Vehicle Company, 26
Walsh, Frank J., 65
Washington Post, 79, 88, 100
Weber, Joe, 2, 37, 72
West, Mollie, 36, 39, 42

women: at funeral of workers, 64; ideals of New Deal and, 21–22; picketing by, 72; protesting at city hall, 69; at rally and march, 37; in SWOC, 5, 72
Woods, Jacob, 81
workers: liberal allies of, 110; movement by and values of, 21, 106–7; risks taken by, 111–12
Works Progress Administration, 63–64

Young, William, 5
Youngstown Sheet and Tube, 1, 24
Yuratovac, Gus, 24, 34

Zwiebach, Peter, 116

www.ingramcontent.com/pod-product-compliance
Lightning Source LLC
Chambersburg PA
CBHW020053170426
43199CB00009B/271